T0332701

# LAYOUT MINIMIZATION OF CMOS CELLS

# THE KLUWER INTERNATIONAL SERIES
# IN ENGINEERING AND COMPUTER SCIENCE

## VLSI, COMPUTER ARCHITECTURE AND
## DIGITAL SIGNAL PROCESSING
*Consulting Editor*
**Jonathan Allen**

**Latest Titles**

# LAYOUT MINIMIZATION
# OF CMOS CELLS

by

**Robert L. Maziasz**
and
**John P. Hayes**

The University of Michigan

**Kluwer Academic Publishers**
Boston/Dordrecht/London

**Distributors for North America:**
Kluwer Academic Publishers
101 Philip Drive
Assinippi Park
Norwell, Massachusetts 02061 USA

**Distributors for all other countries:**
Kluwer Academic Publishers Group
Distribution Centre
Post Office Box 322
3300 AH Dordrecht, THE NETHERLANDS

**Library of Congress Cataloging-in-Publication Data**

Maziasz, Robert L., 1952-
    Layout minimization of CMOS cells / by Robert L. Maziasz and John
P. Hayes.
        p. cm. -- (The Kluwer international series in engineering and
computer science : SECS 160)
    Includes bibliographical references and index.
    ISBN 0-7923-9182-9 (alk. paper)
    1. Metal oxide semiconductors, Complementary--Design and
construction--Data processing. 2. Computer-aided design.
I. Hayes, John P. (John Patrick), 1944- . II. Title.
III. Series.
TK7871.99.M44M33  1992
621.381'52--dc20                                    91-34031
                                                    CIP

**Copyright** 1992 by Kluwer Academic Publishers. Second Printing 2000.

*Printed on acid-free paper.*

Printed in the United States of America

*To our wives*

# TABLE OF CONTENTS

# PREFACE

The layout of an integrated circuit (IC) is the process of assigning geometric shape, size and position to the components (transistors and connections) used in its fabrication. Since the number of components in modern ICs is enormous, computer-aided-design (CAD) programs are required to automate the difficult layout process. Prior CAD methods are inexact or limited in scope, and produce layouts whose area, and consequently manufacturing costs, are larger than necessary. This book addresses the problem of minimizing exactly the layout area of an important class of basic IC structures called CMOS cells.

First, we precisely define the possible goals in area minimization for such cells, namely width and height minimization, with allowance for area-reducing reordering of transistors. We reformulate the layout problem in terms of a graph model and develop new graph-theoretic concepts that completely characterize the fundamental area minimization problems for series-parallel and nonseries-parallel circuits. These concepts lead to practical algorithms that solve all the basic layout minimization problems exactly, both for a single cell and for a one-dimensional array of such cells. Although a few of these layout problems have been solved or partially solved previously, we present here the first complete solutions to all the problems of interest.

We have implemented our algorithms by means of computer programs, and demonstrate that they efficiently generate minimum-area layouts for all possible cells of practical size. An analysis of these layouts permits us to make the first comprehensive evaluation of the quality and scope of prior layout methods. We show that most previous layout optimization algorithms are limited in scope, and cannot find optimal layouts for a large percentage of practical circuits. We also compare our optimal layouts to those of nonoptimal heuristic methods, and demonstrate that our exact algorithms produce significantly smaller cells.

This book will be of interest to designers of circuits or CAD tools, and others concerned with generating highly compact layouts for ICs. Chapter 1 introduces the subject of IC layout, and surveys traditional layout styles. Chapter 2 gives a brief tutorial on the cell layout problem, and evaluates the effectiveness of prior cell layout techniques at minimizing area. Chapter 3 develops our general theory of cell layout and presents algorithms for generating minimum-width cells for series-parallel circuits; Chapter 4 generalizes these results for the nonseries-parallel case. Chapter 5 presents an algorithm that generates minimum-width and -height cells, and these results are extended to an array of cells in Chapter 6. Chapter 7 summarizes the book and suggests applications and extensions to this work.

Most of the material in this book was developed over the past few years as part of the first author's Ph.D. dissertation research at the University of Michigan. This research was supported by grants from the Office of Naval Research and the National Science Foundation, and by fellowships from the Burroughs (now part of Unisys) and Schlumberger Corporations. We wish to express our gratitude to these organizations. We also thank William P. Birmingham, Richard B. Brown, Philip J. Hanlon and Ronald J. Lomax of the University of Michigan for their suggestions.

# LAYOUT MINIMIZATION OF CMOS CELLS

# CHAPTER I

# INTRODUCTION

We begin by discussing the general design problem for integrated circuits, and the role of the cell layout problem in relation to it. Next, we survey the popular cell layout styles for integrated circuits. The cell layout area minimization problem we address is defined and prior work on it is evaluated. We conclude with a discussion of our approach to exact layout optimization.

## 1.1 PROBLEM AND MOTIVATION

Since the introduction of the first commercial integrated circuit (IC) in 1961, the capacity of digital integrated circuits has been increasing at a rapid pace. Commercial IC chips in the 1960's had about 100 transistors. Through dramatic improvements in fabrication technology, very large scale integrated (VLSI) circuits with over ten million transistors are being fabricated at present, and the number of transistors per chip tends to double every year or two. The design of chips with such huge numbers of transistors presents major problems in design effort and cost. Computer-aided design (CAD) tools were developed to help designers cope with these problems. Such tools now play a key role in managing design complexity and improving designer productivity.

The design of a VLSI chip is a complicated process requiring many intermediate steps. Typically, several hierarchical levels of representation are used in the design process; see Fig. 1.1. One of the higher levels is the *register transfer level*, where the variables and the data operators represent the hardware registers and functional blocks of the data-processing section of a chip. The control section implements the sequencing of operations implied by the register transfer description. Next comes the *logic level*, at which the functional blocks are basic logic units called *gates*, such as the AND and NOR gates shown in Fig. 1.1(b); these gates implement the carry function of the adder in Fig. 1.1(a). At the *transistor level*, the logic gates are decomposed into transistor circuits. The NOR gate of Fig. 1.1(b) is translated into the transistor circuit of Fig. 1.1(c), where a *transistor* is a three-terminal device with

1

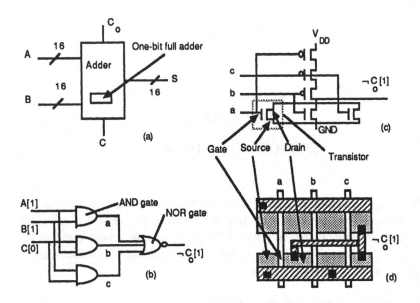

**Fig. 1.1.** Levels of design: (a) register transfer level (16-bit binary adder); (b) Logic level (carry circuit of a 1-bit adder); (c) transistor level (3-input NOR gate); (d) layout level (NOR gate in the CMOS functional cell style).

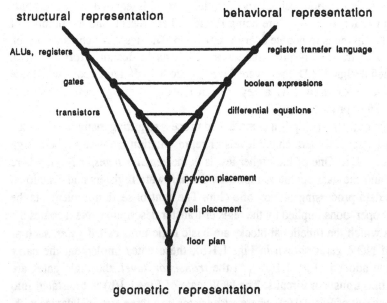

**Fig. 1.2.** Y chart representation of the integrated circuit design process [Ga82].

drain, source and gate terminals. The lowest design level is the *layout level*, where transistors are represented as geometric figures with dimensions such as length, width and position. The geometric representation or *layout* of a basic logic unit is called a *cell*. The layout of Fig. 1.1(d) is a cell implementing the NOR gate in Fig. 1.1(c). A large circuit may be laid out in the form of a set of interconnected cells. As in Fig. 1.1(d), the individual cells of the entire circuit are usually rectangular in shape. The layout design is transferred to mask-making facilities in this form for chip fabrication.

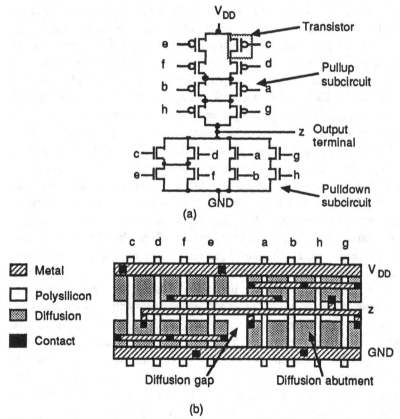

**Fig. 1.3.** A logic function implemented by a functional cell; (a) transistor circuit; (b) layout.

Each of the above design levels has behavioral, structural and geometric aspects. The Y chart in Fig. 1.2 illustrates these aspects, showing the order in which they are considered in a typical hierarchical, top-down design process [Ga85]. At the higher levels, the behavioral aspects are more prominent, and the structural and geometric aspects are less important and less well defined. For example, at the register transfer level, the system is represented as registers and functional units such as an adder. These parts and their interconnection constitute the structure of the system. The behavior may be described by a register transfer language statement, such as

$$register\_A := register\_B + register\_C.$$

Each of the functional units and interconnections at this level can be roughly sized, shaped, placed and oriented in what is called a floor plan [MC80]; see Fig. 1.2. At the transistor level, all structure is explicit; the parts are low-level design components, such as transistors and wires, and the logical behavior of each part is quite simple. At the layout level, the geometric figures composing each transistor and wire are assigned size and shape; each figure represents one of various materials used in fabricating an IC, such as diffusion, polysilicon and metal shown by different shadings in Fig. 1.3(b). Each of these materials occupies a different physical layer of the IC. At the layout level, structure and geometry are most prominent.

Each of the above design steps can be assisted or even be fully automated by CAD tools. Hardware description languages are needed at each level of representation to model the structure, behavior and geometry of the design as it is refined in top-down fashion. *Synthesis tools* are needed to translate the design from a higher level of representation, like the register transfer level, to a lower one, such as the layout level. *Analysis tools* are useful at each level of representation to analyze the quality of the design in terms of correctness and performance. Examples of CAD tools that aid IC designers are logic synthesizers, placement and routing programs, cell generators and simulators. A logic synthesizer can take a design described at the register transfer level and translate it into a set of logic (boolean) equations, accounting to some extent for speed, power and area requirements. These equations can be mapped directly into basic logic units such as AND, OR or NOT gates. A cell generator can produce a cell for each logic unit. The cells can be placed and routed on the chip by placement and routing tools. Simulation at the register transfer, gate and transistor levels can predict the performance of the design at each of these intermediate levels. The designer can use this information to decide whether the behavior of the design meets its specifications.

The *layout style* used in a chip design refers to the rules governing the transistor circuit structure (the circuit type), and the rules for mapping the circuit into the layout (the cell type), and has a profound effect on the amount of circuitry that fits on a chip of a given size. There are two major layout styles, which we discuss in detail in the next section. *Semi-custom* methods select manually designed cells from a cell library; these cells are well characterized in advance for speed and power. In *full-custom* methods, on the other hand, a new type of cell is designed as needed for each logical unit. This can be done by human designers or by automatic cell generators, and requires simulation to analyze the performance of the cell.

This book is concerned with the translation of a design from the transistor circuit level to the layout level. Figure 1.3 shows the main concepts involved in this process. A transistor has three external terminals called the drain, source and gate; Figs. 1.1(c) and (d) show the correspondence between the transistor terminals and the layout. The conducting path between the drain and source terminals is open or closed depending upon the voltage level on the transistor's gate terminal; thus a transistor acts like an on-off switch. The transistor circuit of Fig. 1.3(a) is of a type called complementary MOS or CMOS, and uses two different types of transistors: p-channel MOS transistors in the pullup subcircuit and n-channel MOS transistors in the pulldown subcircuit. Each of these subcircuits has two designated terminals: the pullup subcircuit, whose designated terminals are the logical high voltage, referred to as $V_{DD}$ or 1, and the output $z$; and the pulldown subcircuit, whose designated terminals are the output $z$ and the logical low voltage or ground, denoted as GND or 0. (The terminals to which the input signals $a, b, ..., h$ are applied are not considered designated terminals.) The transistors composing the pullup and pulldown subcircuits create conducting paths connecting the output $z$ either to $V_{DD}$ or GND. Thus $z$ implements a logical function that maps each combination of the input variables onto either 0 or 1.

An important layout task is to place and orient the transistors on the chip to minimize the layout area of a circuit. Figure 1.3(b) shows a cell implementing the transistor circuit of Fig. 1.3(a) in a very area-efficient layout style called a *functional cell* [Uv81]. The various types of shading indicate the materials used to fabricated the transistor terminals and the interconnections between these terminals. The drain and source terminals of a transistor are made from diffusion and the gate terminal is polysilicon. The pullup transistors of the circuit in Fig. 1.3(a) are placed side-by-side in the top diffusion row of the cell of Fig. 1.3(b) such that their drain and source terminals are adjacent, and the pulldown transistors are placed similarly in the lower row. Transistors in the two subcircuits sharing a common gate signal are placed in the same vertical column, which makes for easy routing of the signal between them using a polysilicon column. If the transistors are placed and oriented appropriately, the width of the cell can be reduced by arranging adjacent transistors to share the same drain/source terminal via diffusion abutment, as shown in Fig. 1.3(b). This obviates the need to have separate diffusion terminals which must be isolated by an area-wasting diffusion gap, as also illustrated in Fig. 1.3(b). A judicious placement and orientation of the transistors can result in fewer metal rows for interconnections, thus reducing cell height. Therefore, it is a basic goal of a layout generation tool to lay out the transistors of a circuit and their interconnections to minimize cell area.

We have chosen to address the topic of cell layout generation because it is so important in the overall chip design process. A significant reduction in the chip area required for a given circuit can be achieved by using full-custom layout, and a dramatic increase in productivity over hand layout can be realized by automatically generating these layouts. Layout generators exist for several full-custom layout styles; these styles are surveyed in the next section. Based on the level of recent research activity, the most promising of these styles is the functional cell. However, existing functional cell methods often produce layouts that have far from optimal area. This is due to the fact that most use heuristic or approximate approaches, rather than exact algorithms. An algorithmic layout methodology is needed that achieves minimal area in the functional cell layout style.

Therefore, the general problem being addressed in this book is that of generating minimum-area cell layouts of CMOS digital integrated circuits in the functional cell style. We present a complete solution to the problem of exact layout-area minimization under precisely specified conditions. We will demonstrate that exact area minimization of all practical-sized static CMOS circuits in the functional cell style is both computationally feasible and produces significantly smaller layouts than previous methods. We analyze the properties of the important class of dual planar complementary CMOS circuits, and present efficient algorithms for obtaining the optimal area layout of these circuits. These algorithms, which we have also programmed, are based on a complete theory of cell layout, which significantly extends prior work by Uehara and vanCleemput [Uv81] and others.

## 1.2 LAYOUT STYLES

Digital integrated circuits are often classified according to their layout styles, which are either semi-custom, such as gate arrays and standard cells, or full-custom. In this section we survey both semi-custom methods and the following full-custom styles: unstructured methods, programmable logic arrays, gate matrix methods, and functional cells.

### 1.2.1 Semi-Custom Layouts

Gate arrays constitute a popular approach to the design and layout of digital circuits [Ra80]. A conventional *gate array* consists of rows and columns of cells separated by routing areas for wires interconnecting the gates; see Fig. 1.4(a). These cells are groups of unconnected transistors that can be wired together to form various types of gates, such as AND (*), OR (+), NOT (¬), NAND or NOR. Fig. 1.4(b) shows the structure of a CMOS cell, which has pMOS and nMOS transistors that are interconnected by metal wires (shown by the dotted lines) to form a two-input

NAND gate implementing the function $z = \neg(a * b)$. Typically, a design is represented at the logic level using a library of these basic functions and simulated for correctness and performance. Accurate simulation models can be based on the

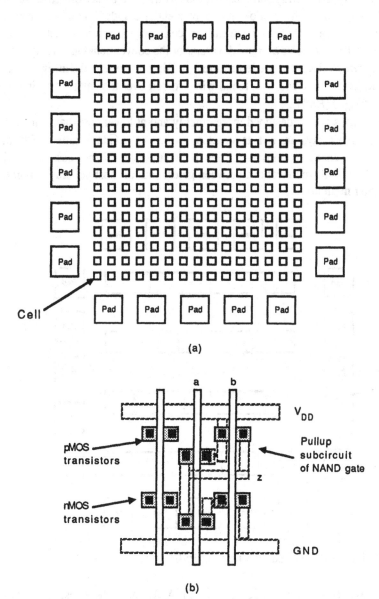

**Fig. 1.4.** A gate array circuit: (a) array of cells; (b) cell structure (after [RU87]).

electrical and performance characteristics of predesigned gates in the library. An alternative to a library of primitive cells is the automatic generation of complex logic circuits that are mapped onto the gate array as needed [Mi86]. Placement of the logic circuits onto the array and their interconnection can be performed automatically by CAD programs. One or more layers of metal are added to the gate array, as in Fig. 1.4(b), to form the gates within the cells and to interconnect these gates according to the particular design. A variant of the conventional gate array is the *sea-of-gates* which has no predetermined routing channels defining cells; gate interconnections are made with metal wires routed over the gates, resulting in more compact designs [Hu85].

Gate arrays can be designed more quickly and at lower cost than standard cells. However, gate arrays are not as small or as fast as standard cells or full-custom designs. They consume more power due to the larger capacitance associated with the typically longer interconnections between gates, and the cost per chip is higher than that of other methods.

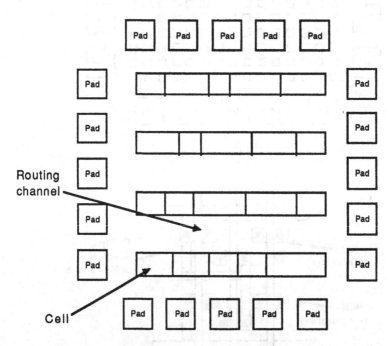

**Fig. 1.5.** A typical standard cell chip layout.

The layouts of basic functions, such as NAND and NOR gates, can be designed by specialists and stored in a library; each such layout is called a "standard" cell. A system designer can then use these standard cells to implement a particular design; this is typically called the standard-cell method. Cells of similar complexity in the cell library typically have a fixed height; see Fig. 1.5. The cells vary in width, depending on their complexity, so they do not lead to the matrix structure characteristic of gate arrays. The inputs and outputs of the cells are on the top and/or bottom of the cells. The cells are placed in horizontal rows on the chip, so that a single horizontal pair of power and ground lines feeds all the cells in a row; Fig. 1.5 shows some typical rows of cells. There is a routing channel between each pair of cell rows which allows interconnection of the cells. Some interconnection is done through vertical spaces left between cells in a row and by connections in the cells. Inter-row connections also use vertical channels along the perimeter of the chip.

The advantages of the standard-cell approach are layouts that are considerably denser than gate arrays and offer higher performance. The greater density is due to the greater compactness of the basic cells and the variability of the routing channel width. This, in turn, results in lower cost per unit. The disadvantage is that the initial design cost is higher because more layers need to be customized than for gate arrays. Also the cost of designing and maintaining a cell library is high.

In addition to semi-custom styles, layout styles have been proposed which need neither a library of standard cells nor an array of configurable cells such as gate arrays. These full-custom styles allow the designer to arrange the transistors and wires in a relatively unrestricted fashion to form cells. There are two types of full-custom methods, which differ in the way they place transistors to construct a cell: unstructured methods and array methods. We survey examples of each type below.

## 1.2.2 Unstructured Methods

*Unstructured* layout methods allow the placement of transistors in any location within a cell. In addition, they allow the wires that interconnect the transistors to be routed in random patterns.

Planar Embedding is an algorithmic method of layout generation which produces nMOS circuit layouts in an unstructured style [NJ85]. Figure 1.6 shows a sample layout produced by this method. Note that the transistors are not aligned to any grid. Using a graph model of the transistor circuit, this method selects a layout that employs the fewest layer changes to interconnect the transistor terminals; in general, such layouts tend to use less area than those requiring more layer changes. However, since the number of layer changes is only indirectly related to layout area, the method does not minimize area and cannot measure how close to optimal its solutions are. The use of such indirect methods of minimizing area is typical of unstructured methods.

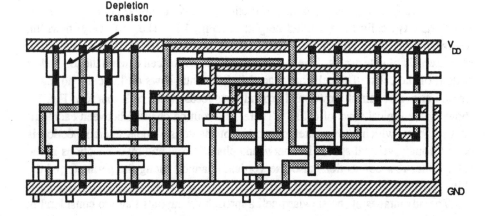

**Fig. 1.6.** A multicell nMOS layout generated by the Planar Embedding technique [NJ85].

The remaining layout styles we survey are called array methods, and place the transistors only at points in an array or grid pattern. These methods are the programmable logic array, the gate matrix, and the functional cell.

## 1.2.3   Programmable Logic Arrays

A *programmable logic array* (PLA) is a widely-used structured layout style [Sm83]. Its logical representation takes the form of a two-level, sum-of-products, multiple-output circuit. Physically, it consists of separate AND and OR arrays, with transistors placed at the grid points of the array; Fig. 1.7 shows the structure of a typical PLA. Each row of the AND array represents a logical product of various input signals, which are the columns of the AND array. These rows extend into the OR array, where the corresponding products are summed. The columns of the OR array then form the logical outputs of the PLA. There are several techniques to reduce the layout area of a PLA [Ul84]. The logic circuit can be transformed to reduce the number of rows in the array. Two inputs can be folded together so that they share the same column in the AND array, thereby reducing the number of columns of the array. Rearranging the order of the rows representing the product terms can increase the number of columns that can be shared. Another technique is the rearrangement of the output columns representing the sum terms to increase the number of rows that can be shared, thus reducing the number of rows of the array.

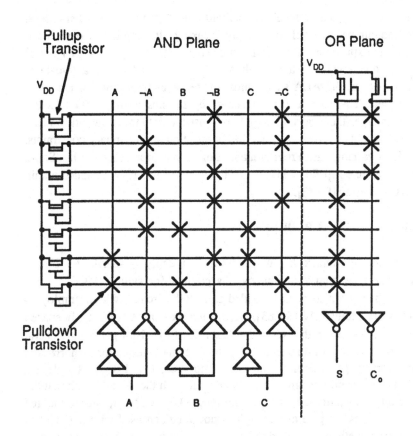

**Fig. 1.7.** A programmable logic array (PLA) implementing a full adder.

The standard PLA structure has several limitations. It does not handle logic functions in nonsum-of-products form; the outputs of the PLA must be routed back to the input to implement such functions, which is wasteful in area and speed. Standard PLA's maintain separate AND and OR arrays; however, the arrays often have few transistors and can be merged into a single array. The long interconnections are quite area-consuming and slow down the circuits.

Modifications of the standard PLA have been devised to remedy some of these limitations. The Associative Logic Array (ALA) proposed by Greer [Gr76] has the following features. The AND and OR arrays are merged into a single array. This allows internal feedback of intermediate functions to be made near the place where they are used as inputs, reducing long feedback paths outside the array. The merging of arrays allows the sharing of rows by two or more product terms generated within the array, by grouping the input columns of the product terms with the OR column to which they connect. Internal feedback facilitates the implementation of *multilevel logic*, where the number of levels is, roughly speaking, the amount of parenthesis nesting in a logic expression of the following style. For example, $z = (a + b + c)$ corresponds to one level of logic, whereas $z = (a + (b * c))$ is a two-level expression. However, the ALA shares most of the other limitations of the PLA. Another modified PLA scheme was presented by Lursinsap and Gajski [LG84, LG85]. It also merges the arrays, folds pullup transistors into the array, allows pass transistors, and accepts a multilevel function description and boundary constraints. Yet another PLA variant, Path Programmable Logic, has been proposed by Smith that allows circuits like flip-flops, inverters, or pass transistors to be included within a merged AND-OR array [Sm83].

## 1.2.4   Gate Matrix Layouts

Next, we turn to matrix layout methods that apply primarily to CMOS technology. The gate matrix layout style is popular for static CMOS circuits [LL80], although it has also been extended to nMOS circuits, which employ n-channel transistors exclusively [SM83]. A *gate matrix* consists of equally-spaced vertical polysilicon columns, that serve the dual role of transistor gates and general interconnections, and rows of diffusion and metal, with the intersection of horizontal diffusion and vertical polysilicon forming transistors; see Fig. 1.8. This style is a generalization of the semi-custom standard-cell style with the modification that the logic and the interconnection channels are no longer kept separate, but are merged together. The full-custom gate matrix style is not to be confused with the similar-sounding gate array method, described in Section 1.2.1 under semi-custom methods.

The number of polysilicon columns in a gate matrix design is typically equal to the sum of the number of distinct inputs and output signals. The diffusions used for transistor terminals are grouped together to form rows. The interconnections required to complete the circuits are made by metal and diffusion lines, which run parallel to, and in between, the polysilicon columns, or by metal along the rows of the matrix. The columns may be ordered to reduce the number of rows. The gate matrix method has been automated using graph-theoretic algorithms which attempt to minimize the number of rows in the matrix by ordering the columns so that the rows can be shared by several gates [Wi82, Wi83, Wi85]. Boundary constraints have also been added to the gate matrix style [YK85]. A variation of the gate matrix style is called metal-oriented layout [Pi83]. It is quite similar to the gate matrix style, with the difference that rows are diffusion and columns are metal instead of polysilicon. This results in less input resistance due to the use of metal input columns. Rows and columns of the gate matrix are folded in the variation of Lin et al. [LD89].

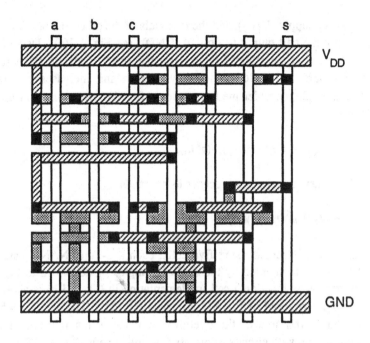

**Fig. 1.8.** A gate matrix layout (after [WE85]).

The gate matrix style eliminates the need for the ground columns and rows used in PLAs and ALAs. It can implement multilevel logic easily because it can route wires between gates inside the matrix. One of the limitations of the gate matrix style is that all the columns are widely spaced to allow for contacts and vertical interconnections. Unless multiple layers of metal are used for interconnection, the long polysilicon gate lines of the standard gate matrix introduce considerable parasitic resistance, and long diffusion wires add much resistance and capacitance, which reduces the circuit speed. Current automatic layout methods are not able to minimize exactly the area of gate matrix designs.

## 1.2.5 Functional Cells

The above full-custom styles often produce large layouts containing many logic gates. In contrast, a functional cell is a layout of a basic logic unit, such as a single gate. A functional cell can implement multilevel logic of arbitrary depth, although this depth is usually limited by performance considerations. We are primarily interested in cells for static CMOS gates, which consist of two 2-terminal subcircuits of transistors, the terminals of one, the pullup subcircuit, connecting the cell output to the power supply ($V_{DD}$), and the terminals of the other, the pulldown subcircuit, connecting the cell output to ground (GND); see Fig. 1.3(a). Using a functional cell to implement a multilevel logic element instead of several standard gates, as in the standard cell style, reduces layout area and increases circuit performance. For example, consider the 8-input logic function represented in the usual infix notation

$$z = \neg((a*b)+((c+d)*(e+f))+(g*h)),$$

or, in prefix notation, where the operator precedes its operands

$$z = \neg(+(*ab)(*(+cd)(+ef))(*gh)).$$

This function can be implemented using five cells, each a 2-input NAND gate. Alternatively, it can be implemented by the single "complex" gate shown in Fig. 1.3(a); the layout area of its functional cell implementation in Fig. 1.3(b) is about half that of the circuit composed of five NAND gates [Uv81]. The use of such functional cells reduces the area of the overall circuit and often improves its performance in comparison with circuits using more primitive cells like NAND and NOR alone. There are two basic types of functional cells, one-dimensional and two-dimensional cells, which we examine next.

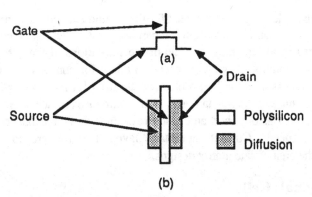

**Fig. 1.9.** An nMOS transistor: (a) circuit symbol; (b) layout.

**One-Dimensional Cells.** A *one-dimensional* functional cell is an array of transistors in which all drain and source terminals of transistors of a given type, either pMOS or nMOS, lie along a single row of diffusion. Figure 1.9 shows the correspondence between the circuit symbol of a transistor and its corresponding layout in a functional cell. A transistor is formed where a polysilicon region crosses over a diffusion region. The polysilicon column of Fig. 1.9(b) defines the gate terminal of the transistor, which determines whether the circuit between the source and drain terminals is open or closed. Thus, the transistor acts as a switch that is on or off depending on the voltage on the gate terminal. The diffusion region on one side of the polysilicon in the figure is the source terminal and the other diffusion region is the drain terminal. Such transistor layouts are placed end-to-end in a horizontal line, with vertical polysilicon columns.

The functional cell style was introduced by Uehara and vanCleemput in 1978 [Uv78] in the context of static CMOS functional cell layout. A static one-dimensional functional cell consists of two parallel linear arrays of this sort, the lower array for the nMOS transistors implementing the pulldown subcircuit of the cell, and the upper one for the pMOS transistors implementing the pullup subcircuit. The gate terminals of transistors sharing the same column are typically connected together by polysilicon. Figure 1.10 illustrates a functional cell of this sort, showing the two diffusion rows, the pMOS and nMOS transistor regions, and the polysilicon columns. In 1985, McMullen and Otten proposed a modified version of this style for dynamic functional cells [MO85], which employ just a single linear array of transistors. We shall examine both static and dynamic one-dimensional functional cell layout styles in the next section.

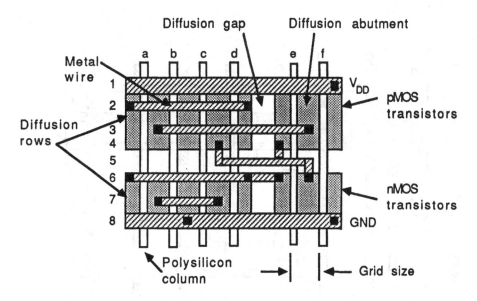

**Fig. 1.10.** A one-dimensional static functional cell showing diffusion regions, metal wires, and transistor types.

**Two-Dimensional Cells.** In contrast to one-dimensional cells, *two-dimensional* functional cells do not restrict the diffusion of the transistors of either the pullup or pulldown subcircuit to a single horizontal line; the diffusion follows the transistor circuit topology. In essence, the circuit is embedded in the two dimensions of a plane using diffusion. Figure. 1.11 illustrates a two-dimensional cell implementing the function $z = \neg(abd + \neg acd + a\neg b\neg c + \neg a\neg bc)$. Typically the gate signals are metal lines which pass over the diffusion without creating transistors, except where the metal is connected to polysilicon transistor gate terminals.

The two-dimensional functional cell style seems to have originated with Weinberger in 1967 [We67]. He proposed a layout style, now called a Weinberger array, which employs a horizontal array of nMOS functional cells with horizontal and vertical pulldown diffusion lines, and horizontal polysilicon rows which serve as the pulldown gate terminals where the two layers intersect.

Other circuit styles have been adapted to the two-dimensional cell style. One such style is cascode voltage switch logic (CVSL), in its dynamic single-ended (SCVS) [Ju90], static single-ended [Er85], and differential (DCVS) [YH85] forms. Other circuit styles implemented as two-dimensional cells are domino CMOS logic [CK89], and static CMOS logic in a related layout style called diffusion strings [TS88].

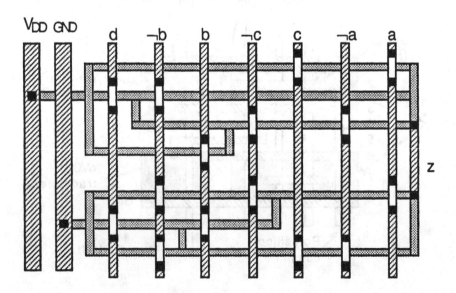

**Fig. 1.11.** A two-dimensional CMOS functional cell.

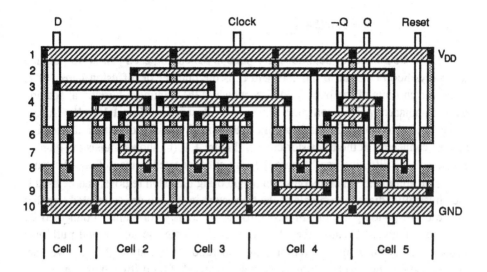

**Fig. 1.12.** A cell array composed of five functional cells forming a master-slave D flip-flop.

**Cell Arrays.** In the functional cell style of layout, logic functions are typically partitioned into multilevel logic circuits, composed of several gates. A functional cell implements each circuit. There are two ways to interconnect functional cells. One is similar to that used in the standard cell style (Section 1.2.1), where the cells are assigned to rows and the cells are interconnected through routing channels separating each pair of rows, as shown in Fig. 1.5. An alternative method groups functional cells to form a *cell array*, which is an array of interconnected functional cells. Figure 1.12 illustrates a *one-dimensional* cell array implementing a master-slave D flip-flop composed of five functional cells. A *two-dimensional* cell array consists of multiple rows of one-dimensional arrays.

## 1.3 FUNCTIONAL CELL OPTIMIZATION

Here we introduce the layout problem for one-dimensional CMOS functional cells. We also summarize and critique prior approaches to solving this problem.

### 1.3.1 Layout Problem

The primary goal in functional cell design is to minimize the cell width W, which is given by the following formula [Uv81]:

$$W = T + G + 1$$

where $T$ is the number of dual transistor-pairs in the circuit, each pair corresponding to a polysilicon column in the cell, and $G$ is the number of diffusion gaps. $W$ is measured in units of the basic grid size, which is the minimum horizontal spacing between adjacent polysilicon gate columns, as illustrated in Fig. 1.10. A diffusion gap is needed to isolate transistor terminals that are physically adjacent in the cell but are not connected in the transistor circuit. Figure 1.10(b) shows transistors $d$ and $e$ separated by a diffusion gap because they are not connected in the pullup subcircuit. Diffusion abutment, which merges two neighboring diffusion regions, can be used if the physically adjacent terminals of the transistors in the cell are connected together in the corresponding circuit. Figure 1.10 shows transistors $e$ and $f$ connected by diffusion abutment because they are connected in both the pullup and pulldown subcircuits. A diffusion gap requires the separation between neighboring vertical polysilicon columns to be twice as large as that needed by diffusion abutment; see Fig. 1.10. Therefore, in the interests of reducing cell width, the preferred method of connecting physically adjacent transistors is diffusion abutment. Thus, the *width minimization problem* to be solved is that of finding a permutation of the polysilicon gate columns and an orientation of the transistors that minimizes $W$ or, equivalently, maximizes the number of transistors that can be connected by abutment.

A secondary design goal is to minimize the cell height, which is determined by the number of horizontal routing rows needed for interconnecting transistor terminals that are not physically adjacent in the cell; such interconnection is done with metal wires as shown in Fig. 1.10. Transistors $a$ and $d$ are connected in the pullup subcircuit, but are not adjacent in the cell of Fig. 1.10; therefore, their terminals are connected by a horizontal metal wire. Figure 1.10 numbers the metal routing rows along the left side of the cell, whose height is shown to be eight rows; that of Fig. 1.12 is ten rows. The *height minimization problem* is to find a permutation of the polysilicon gate columns and an orientation of the transistors that minimize the number of horizontal routing rows required for interconnecting the transistors in the cell. Note that cell width and height are here defined with respect to a grid of polysilicon columns and metal rows. The corresponding width and height minimization problems also assume this grid, and layout sizes are computed with respect to the grid. If a *fixed grid* is used, then final layout size is found directly by scaling the number of grid units by the horizontal and vertical grid sizes. However, if a *virtual grid* is employed, then it may be possible to compact the cell layout by manipulating its components limited only by the layout-level design rules [WE85]; this compaction problem is beyond the scope of this book. Minimizing cell width and height is typically called area minimization, although there may be cases where a minimum area cell has neither minimum width nor height. These functional cell area-reduction goals are in contrast with the goals of gate matrix layout, which primarily attempts to permute transistors to decrease cell height, and typically does not attempt to reduce layout width.

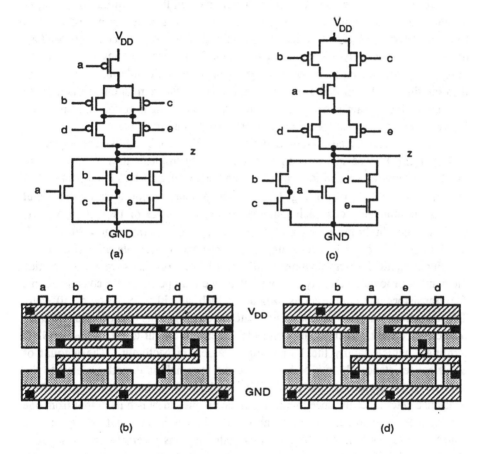

**Fig. 1.13.** (a) Circuit implementing function $z = \neg(+a(*bc)(*de))$ and (b) its layout; (c) Equivalent reordered circuit and (d) its smaller layout.

The transistors of a circuit can often be reordered without changing their logical function. This logical equivalence is based on the commutative property of the logical AND (OR) operation, which is implemented in the circuit by transistor subcircuits connected in series (parallel). Reordering of a circuit can affect its speed and layout area. The impact of reordering on layout area is illustrated by Fig. 1.13(a), which shows a circuit implementing the function $z = \neg(*a(+bc)(+de))$, and an equivalent circuit (c), which reorders the series subcircuits (+*bc*) and *a*. The layout in Fig. 1.13(b), corresponding to the circuit (a) in the figure, is one unit (column) wider than the layout (d) of the reordered circuit (c). Therefore, another area-minimization problem is to find a reordering of the transistor circuit that minimizes the width and height of the functional cell.

These three area optimization problems, width minimization, height minimization, and circuit reordering, are related to one another for a single cell as follows. Width and height minimization are typically the primary and secondary objectives, respectively. A layout of minimum height from among all layouts of minimum width is then the overall design goal. If circuit reordering is performed, then the final goal is to find a layout of minimum height from among all layouts of minimum width for any ordering of the circuit. If reordering is not done, then such a layout is sought only for the original ordering of the circuit.

A large circuit can be composed of multiple cells, each being like the one shown in Fig. 1.13(a). The layout design of a cell array implementing such a circuit involves generating layouts for each of it component cells, such as the layout shown in Fig. 1.13(b) for the cell of Fig. 1.13(a). The problem of coordinating the layout of these component cells such that the composite layout of the cell array has minimum height from among all composite layouts of minimum width for all reorderings of the component circuits is the corresponding *multiple cell* problem.

Functional cells can implement multilevel logic circuits; however, in practice the number of logic levels is typically limited by circuit performance considerations. The transistors in a circuit act as switches, which are either in the on or off state. The transistors that are in the on state create conducting paths between the terminals of their subcircuit. Such a path connects the circuit output to the $V_{DD}$ or GND terminals, which supply the 1 or 0 logic values, respectively. The amount of electrical resistance along a path between the terminals is directly related to the number of transistors on the path, which is defined as the path's length. This resistance determines the speed with which the circuit can switch its output from one logical state to the other. It is typical to limit the maximum-length path between either subcircuit's terminals because this path length is inversely proportional to circuit speed. This maximum length path is called the *height* of the circuit. The height of the circuit in Fig. 1.13(a) is three, since its pullup subcircuit has a path *abd* of length three and none of length four. The upper limit on circuit height defines the class of *practical-sized cells*. A typical limit on the pullup and pulldown subcircuits height is four [Uv81], although a height of up to six may be feasible for the pulldown subcircuit.

## 1.3.2 Prior Work

We now summarize and evaluate the prior work on one-dimensional functional cell layout methods. A more complete survey and critique of the individual methods is given in Chapter 2. There has been much recent research interest in automating the layout of circuits in this style. Most of the methods reported are heuristic in nature, and consequently are inherently nonexact. Such methods claim to find good, but not necessarily minimum-area, layouts. The few nonheuristic methods are based on algorithms of limited scope. They solve parts of the area-minimization problem exactly, but either do not address other parts or solve them heuristically. For example, Hwang et al. solve the width minimization problem but do not deal with circuit reordering and height minimization [HH89]. Nair et al. solve the width minimization and circuit reordering problem for some of the single-cell circuits of interest, but do not address the height minimization and multiple-cell problems [NB85]. Ong et al. present an exact solution to the multiple-cell width minimization problem. However, they do not deal with circuit reordering and solve the cell height minimization problem heuristically [OL89]. Madsen [Ma89] and Lefebvre and Chan [LC89] propose a solution to the width minimization and circuit reordering problems for a portion of the single-cell circuits of interest, but either do not handle the remaining circuits or lay them out heuristically. No prior method solves the width, height, and reordering problems for all single-cell circuits of interest. Furthermore, no prior method solves the width-minimization and circuit-reordering problems for multiple cells, and none solves the height minimization problem for multiple-cell layouts, with or without circuit reordering.

Another characteristic of the prior work is the way the methods are evaluated in terms of layout quality and execution speed. Some researchers report comparisons of a few of their automatically generated layouts to those designed manually. Others compare a relatively small number of their layouts to those published in the research literature. But, in most cases, the evaluations are based on relatively few comparisons; furthermore, the layouts to which comparison is made are not known to be of minimal size. No method is comprehensively evaluated by comparing its layouts to the minimum-sized layouts of all practical-sized circuits. Moreover, only a few papers report the worst-case time complexity of their methods. Some give execution times for a few examples, but times for all practical-sized circuits have not been reported. Since many of these methods may have exponential worst-case time complexity, a few examples may give little insight into their overall performance. (A procedure has *exponential* worst-case time complexity if its execution time grows as a function that is exponential in the size of the procedure's input.)

## 1.4  PROPOSED APPROACH

This section introduces our method of solving the one-dimensional functional cell area-minimization problems defined in Section 1.3.1.

### 1.4.1  Philosophy

We begin by asking the following general questions: under what circumstances should a heuristic be used to solve a hard design problem instead of an algorithm, and if a heuristic is used, what characteristics should it have? Before we discuss the answers to these questions, we define some important concepts.

An *exact algorithm* is a step-by-step procedure for solving a problem $\Pi$ if, given any instance $I$ of $\Pi$, it is guaranteed always to produce an exact solution for that instance $I$ in finite time. For instance, where $\Pi$ is a minimization problem, an exact solution must be of minimum size [GJ79]. A *heuristic algorithm* for a problem $\Pi$, by contrast, does not always find an exact solution for each instance $I$ of $\Pi$.

For a hard problem $\Pi$, a heuristic algorithm may offer better performance guarantees than any known exact algorithm for $\Pi$. However, except in special cases, the quality of a result found by a heuristic algorithm can be arbitrarily far from optimal. In some cases, heuristic algorithms can guarantee near-optimal results [HS78]. A heuristic algorithm $H$ is an *absolute-approximation algorithm* if for every problem instance $I$, its solutions $H(I)$ are never worse that the optimal solution $OPT(I)$ by more than some constant $k$. For example, if $H$ is a heuristic algorithm for a minimization problem, and the solutions are single numbers, $H(I) \leq OPT(I) + k$, for all $I$. A heuristic algorithm is a *relative-approximation algorithm* if, for every instance $I$, $H(I) \leq r \times OPT(I) + k$, for some constants $r$ and $k$. One example of a relative approximation algorithm for wire routing is presented by Brady and Sarrafzadeh [BS90], whose method finds a routing pattern with at most 4/3 times the optimal area. Therefore, if a heuristic algorithm is used, and one wishes to have some guarantee about the distance between the heuristic solution and the exact solution, then either the heuristic solution must be bounded as, for example, in absolute or relative approximation algorithms, or else one must evaluate the heuristic and an exact algorithm on each problem instance in the domain of interest and measure the difference. If none of these evaluations is made, then one must be prepared to accept arbitrarily bad solutions from the heuristic algorithm.

The advantage of an exact algorithm is its guaranteed optimality. Its disadvantage is that its run (execution) time can be very long for large or hard problems. In cases when the problem size is large enough to make the run time unacceptable on the available computers, one is forced to use a heuristic that has an acceptable run time. An exact algorithm is generally considered to be too slow or intractable if its run time on a problem of size $n$ cannot be bounded by a polynomial function of $n$

However, a gray area exists between these two extremes. What should be done when either it is known that no algorithm for $\Pi$ can have polynomial worst-case time complexity, or that no polynomial-time algorithm is known for $\Pi$, and the problem size is bounded at some small or medium value? This is the type of problem that we face in the functional cell layout problem. Most CAD tool designers choose heuristics to solve such problems, whereas we choose to find exact algorithms. It is our philosophy that when this choice is faced, one should first try to find an exact solution until one is convinced that this approach is not feasible; only then should one resort to heuristics. If a heuristic is to be used, an approximation algorithm should be sought. Only if all these attempts fail, should one settle for an unbounded heuristic, which should then be thoroughly tested empirically.

The reasons for this overall approach are threefold. First, the execution time for a small problem size may be quite reasonable. Second, it may also be found that the exact solutions are greatly superior to the heuristic solutions. Third, optimality is theoretically interesting for its own sake, and results in provably minimal cost. We found all of these reasons to be true in our problem of cell layout.

This general philosophy, together with intuition gained from our study of the functional cell layout problem, led us to formulate a working hypothesis at an early stage of our work, which formed the basis for our research. It is as follows.

> **Hypothesis:** Exact solutions to both the single- and multiple-cell width and height minimization problems over all circuit reorderings are both computationally feasible for all practical-sized static CMOS circuits and produce significantly smaller layouts than previous heuristic and limited-scope algorithmic methods.

We follow Uehara and vanCleemput and others in defining the class of all practical-sized static CMOS circuits to be those with a circuit height limit of four [Uv81, LC89, TI86, AT90]. We do not precisely define the limits of computational feasibility, since these limits are changing rapidly with advances in computer technology. However, in general, we consider average execution time on the order of minutes and worst-case time on the order of hours on a small computer, such as a CAD workstation, to be within the limits of feasibility for hard problems such as these. Our goal, stated succinctly, is to demonstrate the validity of the above hypothesis.

We shall reserve the term algorithm for an exact algorithm, and we shall generally use of the terms heuristic for heuristic algorithm. Although the algorithms we present generate layouts within a specific style based upon a set of typical layout assumptions, these algorithms are also quite amenable to other layout styles, such as those that use multiple metal layers for routing.

## 1.4.2  Outline of the Book

The book is organized as follows. Chapter 2 introduces the subject of one-dimensional functional cell layout design. Prior methods are analyzed in detail. We develop a system of classifying all functional cell area-minimization problems, and categorize prior methods according to the problems they address. In Chapter 3, we discuss series-parallel cell-width minimization with and without reordering. We present a comprehensive theory of covering a graph model of dual series-parallel circuits. This theory forms the basis for two layout algorithms, *TrailTrace* and *R-TrailTrace*, which solve exactly the single-cell width-minimization problem for fixed and reordered circuits, respectively. We demonstrate the feasibility of these algorithms by applying them to all single-cell circuits of practical size. In this study, we also thoroughly evaluate other layout methods and analyze the properties of this class of circuits. Our analysis shows that, in the best case, circuit reordering can reduce layout width by more than 20% over optimal-width layout without reordering.

Chapter 4 generalizes the single-cell theory of Chapter 2 to handle all dual planar circuits, which include many nonseries-parallel circuits. We present a layout algorithm, *P-TrailTrace*, which solves exactly the planar cell-width minimization problem. We apply this algorithm to all nonseries-parallel circuits of practical size to demonstrate the computational feasibility of *P-TrailTrace*.

Chapter 5 extends our single-cell theory to incorporate cell-height minimization. We precisely define the height minimization problem, and present an algorithm, *HR-TrailTrace*, which solves exactly the single-cell width and height minimization problem over all circuit reorderings. Again, we apply our algorithm to all practical-sized circuits, demonstrating its feasibility. Our computed results show that height minimization can reduce layout area of a single-cell by more than 80% over width minimization and circuit reordering alone. We also show significant area improvement over most other single-cell layout methods by applying our algorithm to published layouts produced by these methods.

In Chapter 6, we broaden the single-cell theory of Chapter 5 to deal with cell arrays. We define the height minimization problem for cell arrays and present another layout algorithm, *HRM-TrailTrace*, that solves exactly the multiple-cell width and height minimization problem over all circuit reorderings. We apply this algorithm to several published multiple-cell examples and achieve up to 80% area improvement over other heuristic methods. We also apply our algorithm to several large commercial circuits to show its computational feasibility.

Chapter 7 summarizes our results in terms of the eight functional cell area-minimization problems defined in Chapter 2, and shows how these results prove our hypothesis. It also discusses some practical applications and extensions of this work.

# CHAPTER II

# FUNCTIONAL CELL LAYOUT METHODS

In this chapter, we discuss one-dimensional functional cell design in depth and present a detailed survey and evaluation of all important prior methods for generating layouts in this style.

## 2.1 FUNCTIONAL CELL DESIGN

First we discuss the general principles of functional cell design. We begin by introducing the original set of assumptions defining this layout style, followed by an extended illustration of these fundamental ideas. Uehara and vanCleemput originally proposed the style of interest, which is illustrated by Fig. 1.3(b) [Uv78]. To simplify the resulting design problem, they made various assumptions, which we list explicitly in Fig. 2.1.

---

1. Static CMOS technology is used.
2. The pMOS and nMOS subcircuits are series-parallel connections of transistors.
3. The pMOS and nMOS subcircuits are geometric duals of each other.
4. The circuit height $h$ is limited to four for performance reasons.
5. Each cell consists of two horizontal diffusion rows for pMOS and nMOS transistors. This implies that single p- and n-wells can be used.
6. Complementary pMOS and nMOS transistor pairs are vertically aligned. This allows their gate terminals to be interconnected by vertical polysilicon columns without use of crossovers.
7. Transistor drain and source terminals are connected by diffusion if the terminals are physically adjacent in the layout, or by metal if they are not adjacent. Only one metal layer is used.

---

**Fig. 2.1.** The assumptions defining the Uehara-vanCleemput layout style.

Figure 2.2 illustrates the first four assumptions of Fig. 2.1. A static CMOS gate is a transistor circuit composed of two 2-terminal subcircuits of MOS

27

transistors, a pullup subcircuit of pMOS transistors, the designated terminals of which are $V_{DD}$ and the circuit's output $z$, and a pulldown subcircuit of nMOS transistors, whose terminals are $z$ and GND. These subcircuits are functional duals, meaning that when there is a conducting path between the terminals of one subcircuit, there is not such a path between the terminals of the other subcircuit. The circuit of Fig. 2.2 is often referred to as a static CMOS "complex" gate. A pMOS transistor is shown in the pullup subcircuit. The circuit between the source and drain terminals of a pMOS transistor is closed, creating a path for current to flow, if the voltage on the gate terminal is low (at the logical 0 voltage, assuming the positive convention); otherwise the circuit is open, preventing the flow of current. Conversely, the circuit between the source and drain terminals of an nMOS transistor in the pulldown circuit is closed, if the voltage on the gate terminal is high (at the logical 1 voltage); otherwise the circuit is open. For example, if the signals $c$, $d$, $a$ and $g$ are low, then the path through the pMOS transistors with these signals on their gate terminals connects $V_{DD}$ to the output terminal, and the output signal is logical 1. Note that under these conditions, no path from the output $z$ to GND can be closed because the nMOS dual transistors labeled $c$, $d$, $a$ and $g$ open all possible paths to GND. If signals $a$ and $b$ are high, then the output is connected to GND through transistors $a$ and $b$, and all paths from $V_{DD}$ to the output are open resulting in a logical 0 output. The source terminal of an nMOS transistor in Fig. 2.2 is the

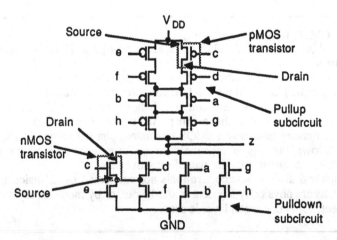

**Fig. 2.2.** A static CMOS complex gate.

one that is closer to GND, whereas the source terminal of a pMOS transistor is the one closer to the $V_{DD}$ terminal; these two cases are illustrated in Fig. 2.2 for the transistors labeled $c$.

The second assumption of Fig. 2.1 is that the pullup and pulldown subcircuits are series-parallel. The circuit of Fig. 2.2 is of this type, where every transistor in either the pullup or pulldown subcircuit is either in series or in parallel with some other transistor. Transistors *a* and *b*, for example, are in series in the pulldown subcircuit and in parallel in the pullup subcircuit. The third assumption is that the pullup and pulldown subcircuits of the circuit are geometric duals [Ha69]. Dual subcircuits reverse the roles of the series and parallel operations; every transistor in series (parallel) with another transistor in one subcircuit is in parallel (series) with its dual transistor in the other subcircuit. Furthermore, dual transistor pairs in the subcircuits have the same signal label on their gate terminals. The pullup subcircuit of Fig. 2.2 the geometric dual of the pulldown subcircuit. The height restriction of the fourth assumption is illustrated by the longest path of transistors *c*, *d*, *a* and *g* in the pullup subcircuit. Since no path in either subcircuit is longer than this, the circuit height is four.

The remaining three assumptions relate to the layout of transistors in the functional cell. First, we briefly discuss the role of the three layers: metal, polysilicon and diffusion. Diffusion and polysilicon are primarily used for creating transistors, which occur where these layers cross, as shown in Fig. 1.9. They can also form interconnections, but they have higher resistance and capacitance than the metal layer, whose primary role is interconnection. The metal layer is on a different plane than either the diffusion or polysilicon layer, and requires a contact to make an electrical connection with these other layers. Two such contacts are shown in Fig. 2.3(b) along the bottom row of metal, interconnecting the diffusion regions of the pulldown regions to GND. Although we only use one metal layer, some IC manufacturing processes offer two or more layers.

The fifth assumption of two rows of transistors is illustrated in Fig. 2.3(b), where the pMOS transistors are placed in the upper row and the nMOS transistors in the lower row. The sixth assumption about interconnecting the gate terminals of a dual transistor pair with a vertical polysilicon column is illustrated in Fig. 2.3(a), where the pMOS and nMOS dual transistors labeled *c* in Fig. 2.2 are vertically aligned in the leftmost column also labeled *c*, and their gate terminals are connected.

Assumption 7 of Fig. 2.1 defines the methods of interconnecting those source and drain terminals in the layout that are connected in the transistor circuit. First, we outline how drain-source terminals of transistors are interconnected or electrically isolated in the functional cell, and how this affects cell size. Terminals connected in the transistor circuit are connected in the cell either by metal wires or diffusion abutment; those not connected in the circuit are isolated either by not placing such terminals adjacent to each other or, if placed adjacent, by inserting a diffusion gap between them. The position and orientation of transistors in the cell determines

which connection or isolation methods are used, and consequently determines the
width and height of the cell.

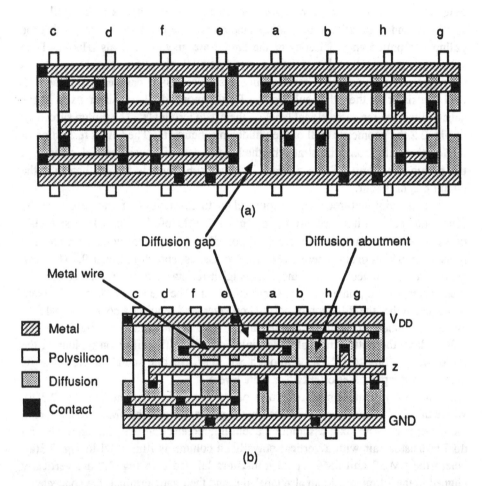

**Fig. 2.3.** A functional cell: (a) with no diffusion abutment; (b) with diffusion
abutment.

The rules for the various interconnection methods allowed by the seventh
assumption are now given. Drain-source terminals that are connected in the
transistor circuit and that are also adjacent in the cell are connected by diffusion
abutment; this case is illustrated in Fig. 2.3(b), where transistors *b* and *h* are
connected in this way. Transistors connected in the circuit but nonadjacent in the

cell are connected by horizontal metal wires; Fig. 2.3(b) shows a metal wire in the pullup subcircuit connecting the diffusion region between columns $d$ and $f$ to the region between columns $a$ and $b$. Drain or source terminals of two transistors not connected in the circuit but adjacent in the cell are isolated by an area-wasting diffusion gap. In Fig. 2.2, transistors $e$ and $a$ are not connected in the pulldown circuit, but are adjacent in the layout of Fig. 2.3(b); therefore, a diffusion gap is required between them.

We illustrate the layout process with Fig. 2.3. Assume we choose to place the dual transistor pairs in the cell in the left-to-right order of $c$, $d$, $f$, $e$, $a$, $b$, $h$ and $g$. Each transistor is *oriented* in one of two ways, such that its drain terminal is either to the left or the right of its polysilicon gate terminal. First, let us consider the cell of Fig. 2.3(a) which uses only the metal layer to interconnect drain-source transistor terminals. A diffusion gap is placed between each pair of transistors, whether the adjacent terminals are to be electrically connected or not. Note that if only metal interconnections were used, all placements of transistors would result in a cell of the same width $W$. In this case, $W = T + G + 1 = 8 + 7 + 1 = 16$ units (columns). However, by choosing a good placement and orientation of the transistor pairs, and using diffusion abutment, the width of the cell can be reduced. Figure. 2.3(b) shows a layout having the same transistor placement and orientation as that in Fig. 2.3(a), but using abutment wherever applicable. The cell width $W = 8 + 1 + 1 = 10$ columns. The height of the cell of Fig. 2.3(b) in terms of the number of metal rows is seven.

The height minimization problem depends on the number of metal interconnection layers used within a cell. Traditionally, one layer of metal is provided for interconnection purposes by chip fabrication processes, although more recently two or more layers have become available. Some layout styles use more than one metal layer for routing within a cell, whereas other styles employ only one metal layer in the cell even when more are available, reserving the other layers of metal for connections between cells. If more than one layer of metal is used within a cell, then in some cases cell height can be made smaller than the height of a cell using only one metal layer. The height problem is also affected by the number of contacts required between a metal layer and the other layers to ensure that a given fabrication process produces a reasonable percentage of working chips and that the total contact resistance is low enough for good circuit performance. For these reasons, multiple contacts become particularly important for submicron processes. The height of a cell may be larger if multiple contacts are required, when compared to a cell using single contacts. The vast majority of prior layout methods use a single layer of metal for cell interconnection; moreover, almost all of them use a single contact between metal layers and the other layers.

As stated in Chapter 1, there are several fundamental area-minimization goals. The primary design goal in one-dimensional functional cell layout is to minimize the cell width by ordering and orienting the transistors in the cell so that the number of neighboring transistors connected by diffusion abutment is maximized, thus minimizing the number of diffusion gaps. Likewise, the secondary goal is to order and orient the transistors in the cell to minimize cell height by minimizing the number of horizontal metal rows needed to connect the nonadjacent drain-source terminals by metal wires. The third goal is to reorder the circuit to reduce both cell width and height. A fourth goal is to solve the width, height and reordering problems for arrays of cells. We define eight layout problems based on which combinations of these four fundamental design goals are addressed; these problems are listed in Fig. 2.4. The names of the layout problems in Fig. 2.4 correspond to the optimization goals the problem address, where W, H, R and M refer to the width, height, reordering and multiple-cell design goals. For example, the WHR problem addresses the width, height and reordering problems, but not the multiple-cell problem. The relationship between these design goals is discussed in Section 1.3.1.

| Layout problem | Width minimization | Height minimization | Circuit reordering | Number of cells |
|---|---|---|---|---|
| W | Yes | No | No | Single |
| WR | Yes | No | Yes | Single |
| WH | Yes | Yes | No | Single |
| WHR | Yes | Yes | Yes | Single |
| WM | Yes | No | No | Multiple |
| WRM | Yes | No | Yes | Multiple |
| WHM | Yes | Yes | No | Multiple |
| WHRM | Yes | Yes | Yes | Multiple |

**Fig. 2.4.** Classification of area-minimization problems.

It has been shown that the WH problem is NP-hard for series-parallel dynamic circuits that possess an euler trail [Ch91]. It has also been noted that the version of this problem for static circuits is "considerably harder" than for dynamic circuits. Therefore, the WH problem for all dual series-parallel circuits is no less than NP-hard, as is true also of the WHR, WHM and WHRM problems for this broader class of circuits.

## 2.2 SURVEY OF PRIOR METHODS

We survey both the static and dynamic one-dimensional functional cell design methods, and classify them according to which of the layout problems of Fig. 2.4 they address.

### 2.2.1 Static Cells

First, we examine the design methods for static functional cells, which have two diffusion rows to accommodate the equal number of transistors contained in the pullup and pulldown subcircuits.

Uehara and vanCleemput originally proposed a method for generating functional cells in the one-dimensional functional cell style [Uv78], based on the list of assumptions in Fig. 2.1. Not all methods generating layouts in the one-dimensional functional cell style make all these assumptions; some methods use them as guidelines, but do not adhere to them consistently. However, most of them are adopted by the majority of layout systems; we will note the methods that make different assumptions.

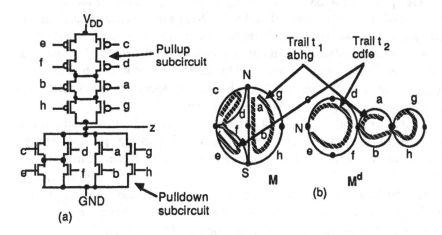

**Fig. 2.5.** The graph approach of Uehara and vanCleemput: (a) transistor circuit; (b) graph model.

We now examine the Uehara-vanCleemput layout method, which is a heuristic for solving the WR problem. The input to the design process is a transistor circuit of the type represented by Fig. 2.5(a). The transistor circuit is modeled by a graph, as in Fig. 2.5(b) which represents the circuit of Fig. 2.5(a). The edges correspond to transistors in the circuit and nodes correspond to interconnections of transistor source and drain terminals. An edge is labeled with the name of the input signal applied to its gate terminal. The graph model consists of two series-parallel multigraphs, $M$ for the nMOS pulldown subcircuit and $M^d$ for the pMOS pullup subcircuit. (A *multigraph* is a graph where more than one edge joins two nodes [Ha69].) The pMOS graph $M^d$ is the geometric dual [Ha69] of the nMOS graph $M$, and vice versa.

The process of translating the graph model of the transistor circuit into a layout involves finding a set of trails in the graphs such that every edge of the graph model is included exactly once in some trail. (A *trail* is an alternating sequence of nodes and edges, beginning and ending with nodes, where no edge is repeated [Ha69].) Each trail in this set can be laid out as a continuous region of diffusion. For example, the trail $t_1 = abhg$ and the trail $t_2 = cdfe$ in Fig. 2.5(b) can be laid out as two diffusion regions separated from each other by a diffusion gap, as shown in Fig. 2.3(b). Therefore, the number of trails must be minimized in order to minimize the layout width.

A key assumption (number 6 of Fig. 2.1) of the Uehara-vanCleemput layout style is that the gate terminals of each nMOS and pMOS transistor-pair corresponding to dual edges in the circuit are connected by vertical polysilicon lines in the layout. This assumption constrains the type of trail that is traced through the graphs; each pair of trails, one in the pMOS graph and the other in the nMOS graph, must consist of the same sequence of dual edges. Such a trail is called a *dual trail* or *d-trail* The trails $t_1$ and $t_2$ in Fig. 2.5(b) are both dual trails, whereas $cdg$ is a trail in $M$ but is not a dual trail. A dual trail that includes all the edges of a graph is a *dual-euler* or *d-euler* trail, as illustrated by the trail $cbaed$ in the circuit of Fig. 1.13(c). As Fig. 1.13(d) demonstrates, a layout corresponding to a d-euler trail contains no diffusion gaps and so has minimum width.

The circuit of Fig. 2.5 does not contain a d-euler trail. In such cases, the Uehara-vanCleemput method modifies the graph model by adding "pseudo-edges" to ensure that the modified graph will have a d-euler trail. It then rearranges the subgraphs of the modified graph in a way that preserves the logical function represented by the graph, and minimizes the number of contiguous groups of pseudo- and real edges along the d-euler trail. This d-euler trail is the basis for the layout, where real edges correspond to transistors and each contiguous group of pseudo-edges corresponds to a diffusion gap. The heuristic does not guarantee a minimal-width layout. When applied to the circuit of Fig. 2.3(a) it finds two dual trails, resulting in the layout of Fig. 2.3(b), which has a diffusion gap. We will show in Chapter 3 that our method finds the optimal layout, which in this case has no gaps.

**The W Problem.** Nair and his colleagues [NB85] have extended the Uehara-vanCleemput approach. Using the same layout assumptions, they have developed an optimal algorithm solving the W (width only) minimization problem, and an algorithm that solves the WR (width minimization with circuit reordering) problem for a subset of practical circuits. The first approach finds the minimum number of dual trails in a graph model of the circuit where the order of the subgraphs is fixed. It reduces sets of dual trails found in series-parallel multigraphs to 18 "representative" graphs based upon the number of distinct trails in the graph and the terminals of the graph that are the endpoints of the trails. These representative graphs are used to keep track of the different trails as the graph is processed. Representative graphs are combined together, resulting ultimately in an optimal set of dual trails. This set of trails leads to a minimum-width layout for fixed graphs. Nair et al. have also developed an algorithm for reordering the subgraphs of the graph model. This method produces optimal results in the special case that some ordering of the graph has a d-euler trail; otherwise it yields no result. We show in Chapter 3 that only 56% of practical-sized static circuits have this property. Both methods have a time complexity that is linear in the number of transistors.

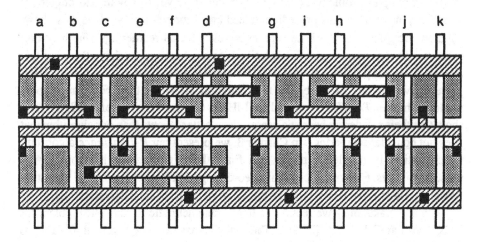

**Fig. 2.6.** A layout of the function $\neg(+(*hi)gjk(*ef)(*d(+c(*ab))))$ by the method of [HS88].

Huang and Sarrafzadeh have generalized the preliminary version of our dual trail-tracing serial algorithm reported in [MH87], by developing a parallel version based on our trail-covering theory [HS88]. It solves the W problem for all practical-sized series-parallel circuits. The time complexity of the method is O(log $n$) time using O($n$) processors. An example layout of minimum width produced by the algorithm of Huang and Sarrafzadeh is shown in Fig. 2.6.

**The WR Problem.** As discussed above, Uehara and vanCleemput were first to address the WR problem. More recently, Kwon and Kyung have reported a heuristic for the WR problem [KK88]. It yields a "nearly minimum" trail cover in $O(n^3)$ time, where $n$ is the number of transistors in the circuit. They compare two layouts of their heuristic to those of our algorithm [MH87], but do not find any area improvement, although they do show improvement over the method of [Uv81]. The strength of this method is that it recognizes the importance of reordering and attempts to develop the best polynomial-time method that reorders the circuit. The authors predict that optimal reordering has complexity $O(n^n)$, whereas we show that its complexity is $O(h!n)$, where $h$ is the height of the circuit, typically a very small number for practical-sized circuits.

Chen and Hou describe a heuristic for reordering and laying out a functional cell [CH88] to reduce its width, but not its height. They replace the original problem of finding the minimum number of trails to cover the dual graph model with the following simpler problem: minimize the number of vertices with odd degree by reordering the graph. This procedure indeed can improve the layout, as they show, but there are graphs all of whose vertices are even for all reorderings. Furthermore, some reorderings may require fewer trails to cover these graphs than others do; this method does not provide any guidance for reordering such graphs. Chen and Hou present five layouts produced by their method to which they compare layouts produced by our *TrailTrace* algorithm [MH87] and the heuristic of [Uv81]. They show that their reordering heuristic can produce better layouts than our algorithm that does not consider reordering. In Chapter 3, we present our *R-TrailTrace* algorithm, which optimally reorders circuits, and finds the minimum-width layout for all practical circuits, 61% of which require reordering. The earlier *TrailTrace* algorithm was designed to solve the W problem only, and so does not consider reordering.

Lin and Nakajima have reported a linear-time heuristic for the WR problem for dual series-parallel circuits [LN90]. Their method has generated optimal solutions to seven examples taken from functional cell layout papers.

**The WHR Problem.** Lefebvre and Chan have developed a heuristic method that appears to solve the WHR problem for a subset of practical-sized circuits [LC89]. The authors adopt the typical practicality constraint that the height of the circuit graph must be four or less. However, they also impose another constraint that the maximum number of logic levels must be three or less. We analyzed this constraint and found that only 18% of the practical-sized circuits meet it. The authors do lift assumption 6 of Fig. 2.1, allowing them to find very small layouts in special cases. In a second paper [LC90], Lefebvre presents a heuristic multicell layout method.

Madsen reports a method for functional cell synthesis that appears to solve the WHR problem under certain circumstances [Ma89]. Madsen's method also produces layouts that meet certain boundary constraints on the ordering of their inputs, and uses reordering to improve the layout area to meet boundary constraints and to improve circuit speed. He finds a d-euler trail in a circuit if one exists. However, he does not claim to find the optimal layout if a circuit cannot be covered by a single dual trail. This method makes essentially the same claim as the second algorithm in [NB85]. As we show in Chapter 3, 44% of all practical circuits do not have a d-euler trail in any reordering. Therefore, neither method cited can guarantee optimality for almost half of these circuits. It is unclear whether the Madsen method always achieves optimal height on circuits with d-euler trails. The author compares his method to those of [Uv81] and [CH88], showing his method yields better results. The layouts Madsen generates using the method of [Uv81] are better than those using the method of [CH88]; however, in [CH88], Chen and Hou show their method to be better than that of [Uv81] on all five examples! This demonstrates the inconclusive nature of evaluating a method based on a few comparisons.

Mailhot and DeMicheli have developed a linear-time heuristic method that attempts to solve the WHR problem [MD88]. The method applies to layout styles using one or two layers of metal interconnection and styles using single or multiple contacts, and can perform some width-versus-height tradeoffs. The method uses a few rules to perform limited reordering of the circuit. The authors present four sample layouts which they compare to those produced by other methods, but are not able to find the optimal layout for the one example taken from our paper [MH87].

**The WM Problem.** Hwang et al. present an optimal algorithm for minimizing the layout width of fixed multicell circuits with equal numbers of pMOS and nMOS transistors, thus solving the WM problem [HH89]. They model the circuit as a bipartite graph and use a branch-and-bound method, which is inherently an optimal technique. The method provides a tight lower bound on the number of trails in the optimal cover. With respect to series-parallel circuits, the results of Hwang et al. are identical to those previously produced by our first algorithm *TrailTrace*, except that our method's time complexity is linear in the number of transistors, whereas their method, which uses a branch-and-bound approach, has exponential-time complexity in the worst case. A second paper presents their method augmented by heuristics to handle cell height [HH90]; this method allows the use of multiple contacts.

**The WHM Problem.** Ong et al. describe a multicell layout system called GENAC that handles general CMOS circuits [OL89]. GENAC minimizes the layout width and, as a secondary optimization goal, it reduces routing channel density and consequently cell height; it does not reorder the circuit graph. The authors do not claim that the resulting cell is of minimum height; therefore, their method addresses, but does not solve, the WHM problem. GENAC starts with pairing pMOS and nMOS transistors, followed by a hierarchical partitioning of these pairs into groups, such that only pairs within a group can potentially abut. For each such group, diffusion line-tracing is done. GENAC combines these separate diffusion lines to form a minimum-width layout. Diffusion lines are abutted so as to reduce routing channel density. Large transistors are folded, and finally the cell is routed. To evaluate the quality of the layout method in terms of area, Ong et al. have synthesized the 72 flip-flop cells in the AT&T standard cell library and compared GENAC's designs to the library cells. They show that GENAC's cells are slightly smaller in terms of width. However, GENAC's cells are usually one or two rows higher than the library cells, which are proportionally more important than diffusion gaps. Ong et al. admit that they are not able to generate all cells with equal or better area than those of the manually designed library cells. Their method is fast, taking only a few seconds to generate small cells on a medium-sized computer. Ong et al. take a mathematically rigorous approach to cell synthesis; they make some precise claims and prove them. They also generalize some important constraints, allowing general circuits, individually-sized transistors, and routing over diffusion. They give no worst-case time complexity for their method, but it seems to be exponential.

Wimer et al. report a functional cell layout algorithm that handles general CMOS circuits [Wi87], and minimizes the number of diffusion gaps for limited-size circuits. However, their method does not reorder the circuit graph and does not minimize layout height; therefore, it addresses the WHM problem. The steps of the method are these. First, p- and n-channel transistors are paired to share a vertical column. Diffusion chains are formed in an exhaustive fashion and assigned a cost reflecting routing density and performance. A set of chains is selected that minimizes the overall cost of the circuit; these chains are placed horizontally in the row so as to minimize interchain routing density. All $2^n n!$ permutations and orientations are generated if the number of chains is less than five; otherwise, a random sampling of all possible arrangements is done. The chains are then routed using a greedy channel routing heuristic. The layout program of Wimer et al. was developed for relatively small circuits with a few dozen transistors, for which it is reported to run "instantaneously." Although this method minimizes the number of diffusion gaps on small circuits, the cell width is not necessarily optimal, since the channel router may introduce additional columns into the layout that increase the cell width.

Bar-Yehuda et al. have presented a method for generating general CMOS circuits in a functional cell layout style with some additional constraints [Ba89]. Their method partially solves the WHM problem for layouts of small height. The layout style has predefined routing channels with three horizontal tracks, one each for the n- and p-channel devices (two for transistor interconnections and one for power routing) and fixed-spacing polysilicon columns. The method minimizes the layout width for fixed circuits only; the circuit graph is not reordered. Layout height is the same for each layout, as determined by the layout constraints. The cell height constraints seem to be needed to ensure the computationally feasible of the method. The steps of the method are the following. The placement of transistors is generated in a depth-first branch-and-bound fashion, using as a bounding function some combination of (in order of importance) diffusion abutment, vertical alignment of polysilicon gates of transistors, and total wire length. A linear-time dynamic programming algorithm orients the transistors to minimize the number of diffusion gaps in the partial chain already generated. After the placement is complete, these chains are routed using a dynamic programming method that is exponential in the number of routing tracks. The method "reaches its limit" in the placement phase, so that larger circuits require partitioning. The method has been run on 300 cells of a manually generated library and compared to them. The manual cells were never smaller and often larger. The algorithm is of exponential-time complexity in the worst case.

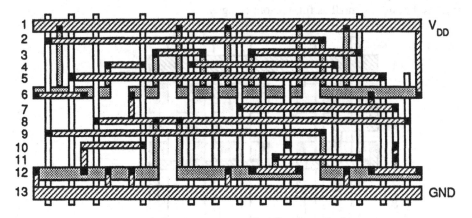

**Fig. 2.7.** A master-slave D flip-flop laid out by TOPOLOGIZER (after [KW85]).

Kollaritsch and Weste developed a heuristic program called TOPOLOGIZER for the layout of general multicell CMOS circuits [KW85]. They use the rule-based paradigm, which attempts to exploit the knowledge of expert designers as built-in design rules. TOPOLOGIZER allows the user to specify boundary constraints like aspect ratio and pin location, and can generate two or more rows of pMOS and nMOS transistors. The placement of transistors is done using rules that iteratively improve an initial placement by pairwise exchange. The object is to maximize the number of short single-layer routing connections. The results from TOPOLOGIZER range from 8% better to 82% worse than production-quality manual layouts based on four comparisons. Rule-based methods are not generally amenable to precise analysis; optimality claims are almost impossible to verify. Therefore, the only way to evaluate the quality of rule-based methods is to compare the area of their layouts to those of known quality. Figure 2.7 shows the only multicell example the authors present in [KW85], a master-slave D flip-flop design, which is 3% smaller than the corresponding manual layout. In Chapter 6, we show that our optimal method produces a layout that is 39% smaller than the layout in Fig. 2.7.

Hill presents a heuristic method called Sc2 for the layout generation of arbitrary circuits in a functional cell style that uses single or multiple contacts [Hi85]. First, it splits tall transistors into shorter ones. It then pairs the pMOS and nMOS transistors and subsequently treats them as vertical neighbors. Next, it horizontally orders these pairs within a row. The transistor pairs are then oriented to increase diffusion abutment using a branch-and-bound algorithm. A heuristic method is employed for intercell routing. Sc2 produces layouts about as small as standard-cell layouts, but not as small as full-custom designs.

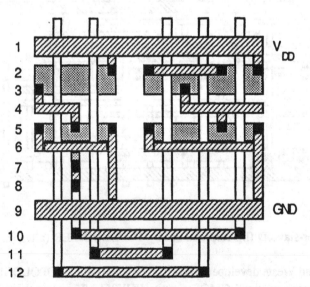

**Fig. 2.8.** An exclusive-OR cell generated by SOLO (after [BA88]).

Baltus and Allen report a heuristic layout method called SOLO that produces layouts in the functional cell style [BA88]. Figure 2.8 shows a layout of an exclusive-OR gate produced by SOLO. Its most distinguishing feature is that it is more general than most of the other functional cell methods in that it can handle nonseries-parallel circuits, transmission gates, boundary constraints, and multiple functional cells. SOLO partitions a circuit into functional cells and other closely related transistor groupings. It uses general placement and routing methods. The resulting method is indeed very general. However, such generality is not without its price; a large penalty in area is paid, as we shall show in Chapter 6, where we present a layout generated by our algorithm with 80% less area than that of Fig. 2.8.

Shiraishi et al. report a heuristic layout system called MAGICAL that also handles arbitrary CMOS circuits [Sh88]. This method uses two layers of metal. Its main goals are, in order of importance, to maximize the number of vertical transistor gate connections, to maximize diffusion abutment and to minimize total wiring length; thus it addresses the WHM problem. The main steps of the method are as follows. First, a connection graph is created that represents all the connections between terminals of the transistors. Diffusion "islands" are created from the transistors: series diffusion islands and parallel diffusion islands from those transistors strictly in series or in parallel, respectively. Next, the islands of nMOS transistors are paired with those of pMOS with which they are connected to form a diffusion island couple. The transistor pairs are placed in a row so as to minimize total wiring length; they are also rotated to maximize diffusion abutment between pairs, if this rotation does not increase wiring length. Pairs that are part of the same functional cell are recognized and placed together. Routing is done in the five channels defined for the cell. Two example layouts presented in [Sh88] are slightly smaller than hand designs; however, the layouts produced are not necessarily optimal in height, width or area.

Chen and Chow present a layout heuristic for general CMOS circuits that utilizes two layers of metal and multiple contacts. It simultaneously attempts to maximize diffusion abutment and minimize wiring area [CC89]. First, transistors are paired based on a hierarchy of priorities, the most important of which is sharing a polysilicon gate column. Next, transistor pair placement is done guided by a set of graph models. A diffusion-sharing graph is transformed into an adjacency graph. The method finds all hamiltonian paths in the adjacency graph by a branch-and-bound algorithm, evaluating each for its area cost. The transistor pairs are then optimally rotated in linear time. Finally, routing is performed. The authors compare two manual layouts having 18 transistors each to those of their method. The hand-crafted design is 8% larger in one example and 1% smaller in the other.

Sun presents a heuristic layout system for high performance circuits [Su89], that does not reorder the circuit graph. This method employs multiple contacts, and two metal layers. Sun's method consists of the following steps. First, pMOS and nMOS transistors are paired. Diffusion chains are created, followed by linear ordering of the chains in a row. Finally, some transistors are placed in under-utilized routing areas. The method's synthesized layouts range from 17% smaller to 53% larger than the ten manual examples presented.

Poirier has developed a layout generator called Excellerator that handles arbitrary CMOS circuits [Po89]. This method uses two levels of metal for interconnection. The optimization goals are to maximize the number of transistors sharing a gate signal, to maximize diffusion abutment, and to minimize routing density and wire length. Although the method appears to produce very compact layouts, it does not guarantee minimal width or height. Excellerator's layouts are from 25% smaller to 6% larger than the manual layouts to which it is compared. The reported run times range from 22 seconds to over an hour on a medium-sized computer for circuits with up to 40 transistors.

**The WHRM Problem.** Mano et al. [Ma85] report a layout system that partitions an input logic expression into functional cells and registers. It uses the technique of [Uv81] to generate functional cell layouts, and places and routes the cells in a standard cell layout style. We compare a layout produced by the method of Mano et al. to one produced by our optimal method in Chapter 6.

Heeb has proposed a functional cell layout system [He87], which extends the heuristic in [Uv81]. He extends the latter method to handle intracell routing of the source and drain terminals of minimum-sized transistors using only metal wires. This involves solving the single-row routing problem for functional cells [Le90]. The goal of the method is to generate a layout that can be routed only in metal and is close to optimal width. Heeb's heuristic, like that of [Uv81], does not find optimal-width layouts in a significant number of cases; the reasons for this will be discussed fully in Chapter 3. In a related paper, Heeb and Fichtner present a heuristic multicell layout scheme that uses multiple rows of functional cells [HF87], based upon their enhanced version of the Uehara-vanCleemput heuristic method.

Domic et al. describe a heuristic layout generator CLEO for arbitrary CMOS circuits [Do89]. CLEO produces layouts of one or more horizontal rows of vertically aligned transistors. This method employs multiple contacts. Its layouts are of comparable area to hand-crafted designs, and it reorders the circuit graph under unspecified circumstances. Thus CLEO attempts to solve the WHRM problem. The method may be summarized as follows. First, the circuits are placed in a given row using a recursive min-cut method [Le90]. A cell layout for each circuit is obtained next. A predefined library cell is used for standard logic circuits and registers, and a layout is generated for functional cells. The individual cells are manipulated to maximize the amount of diffusion abutment by selecting the best cell and orienting it to match its neighbors; gate inputs are permuted to avoid wire crossing when routing cells. A simulated annealing algorithm [Le90] is used for this stage. The next stage involves the formation of diffusion chains by merging transistors via diffusion abutment. The chains are placed horizontally in their rows to minimize signal routing distance. Special-purpose routers that handle two layers of metal are used. CLEO processes about 75 transistors per minute for production circuits. The goals of this tool are generality and flexibility along with small area, which it reportedly achieves.

## 2.2.2 Dynamic Cells

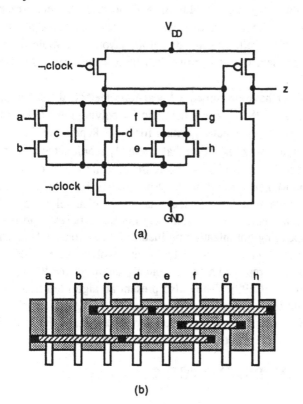

**Fig. 2.9.** A one-dimensional domino-logic dynamic CMOS cell: (a) transistor circuit; (b) layout of the pulldown subcircuit (after [MO88]).

Now we briefly review layout methods for one-dimensional functional cells that employ dynamic CMOS circuits. As mentioned in Section 1.2.5, dynamic circuits replace the pullup subcircuit by a fixed circuit controlled by a clock; therefore, the cell function is determined by the pulldown subcircuit. Consequently, the main focus is on layout of the pulldown circuit, which typically consists of a single row of transistors connected by diffusion. The domino logic circuit in Fig. 2.9(a) is an example of a dynamic circuit realizing the function $z = (a * b) + c + d + ((f + g) * (e + h))$.

McMullen and Otten present an algorithm for generating the layout of the pulldown MOS subcircuit of a dynamic series-parallel CMOS circuit [MO88]. This method reorders the circuit graph to minimize the diffusion gaps, but does not handle cell height; therefore, it solves the dynamic case of the W problem. The style of layout is a one-dimensional transistor cell consisting of a horizontal row of diffusion, with vertical polysilicon columns and horizontal metal tracks, as illustrated in Fig. 2.9(b). The authors propose a linear-time layout algorithm that guarantees a minimum-width cell. In generating 1500 layouts, McMullen and Otten found that only two of them required more than three metal tracks for intracell routing.

Lengauer and Müller have presented a linear-time method for series-parallel dynamic CMOS circuits [LM88]. It is a heuristic for solving the WHR problem, but can also be classified as an exact algorithm for the WR problem. The authors present a dynamic programming algorithm that optimally reorders the circuit graph to minimize cell width; it is similar to the reordering algorithm of Nair et al. [NB85]. Lengauer and Müller do not describe their heuristic for reducing the height of their layout, but state that "in applications where channel density is the dominating issue such a (heuristic) approach may not be sufficient, and the exact solution to the corresponding optimization problem may be indicated." In Chapter 5, we present an exact solution for the related problem involving static CMOS cells.

Chakravarty et al. have reported a linear-time algorithm for the W problem for arbitrary dynamic circuits [Ch90]. They also present an algorithm that finds the minimum-height layout for a given linear placement of transistors. Carlson et al. present an algorithm for solving the WR problem for series-parallel dynamic circuits whose time complexity is linear in the number of transistors [Ca90].

## 2.3  CRITIQUE OF PRIOR WORK

We now evaluate and categorize the prior layout methods according to the area-minimization problems they address, in order to identify the previously unsolved or only partially solved problems; this provides motivation for our research. We conclude by discussing two trends in the layout literature that tend to discourage the search for exact solutions.

**Optimality of Prior Methods.**  Figure 2.10 summarizes prior layout methods in terms of the problem classification of Fig. 2.4. Most of the cited methods use heuristics that do not guarantee a solution of minimum width or height.

| Reference | Layout problem | Width minimization | Height minimization | Circuit reordering | Number of cells | Time complexity | Circuit type |
|---|---|---|---|---|---|---|---|
| [NB85] | W | Exact | None | No | Single | $O(n)$ | SP |
| [HS88] | W | Exact | None | No | Single | $O(\log n)$ | SP |
| [NB85] | WR | Exact[1] | None | Yes | Single | $O(n)$ | SP |
| [CC87] | WR | Exact[1] | None | Yes | Single | ? | Any |
| [Uv81] | WR | Not exact | None | Yes | Single | $O(n)$ | SP |
| [KK88] | WR | Not exact | None | Yes | Single | $O(n^3)$ | SP |
| [CH88] | WR | Not exact | None | Yes | Single | ? | SP |
| [LN90] | WR | Not exact | None | Yes | Single | $O(n)$ | SP |
| [NH85] | WH | Not exact | Not exact | No | Single | ? | Any |
| [CC89] | WH | Not exact | Not exact | No | Single | ? | Any |
| [Su89] | WH | Not exact | Not exact | No | Single | ? | Any |
| [LC89] | WHR | Exact[2] | Exact ? | Yes | Single | ? | SP |
| [Ma89] | WHR | Exact[1] | Exact ? | Yes | Single | ? | SP |
| [MD88] | WHR | Not exact | Not exact | Yes | Single | $O(n)$ | SP |
| [HH89] | WM | Exact | None | No | Multiple | Expon. | Any |
| [OL89] | WHM | Exact | Not exact | No | Multiple | Expon. | Any |
| [HH90] | WHM | Exact | Not exact | No | Multiple | Expon. | Any |
| [Wi87] | WHM | Exact ? | Not exact | No | Multiple | ? | Any |
| [Ba89] | WHM | Exact[3] | Fixed | No | Multiple | Expon. | Any |
| [KW85] | WHM | Not exact | Not exact | No | Multiple | ? | Any |
| [Hi85] | WHM | Not exact | Not exact | No | Multiple | ? | Any |
| [BA88] | WHM | Not exact | Not exact | No | Multiple | ? | Any |
| [Sh88] | WHM | Not exact | Not exact | No | Multiple | ? | Any |
| [Po89] | WHM | Not exact | Not exact | No | Multiple | Expon. | Any |
| [Ma85] | WHRM | Not exact | Not exact | Yes | Multiple | ? | Any |
| [HF87] | WHRM | Not exact | Not exact | Yes | Multiple | ? | Any |
| [Do89] | WHRM | Not exact | Not exact | Limited | Multiple | ? | Any |
| [LC90] | WHRM | Not Exact | Not exact | Yes | Multiple | ? | Any? |

(a)

| Reference | Layout problem | Width minimization | Height minimization | Circuit reordering | Number of cells | Time complexity | Circuit type |
|---|---|---|---|---|---|---|---|
| [Ch90] | W | Exact | None | No | Single | $O(n)$ | Any |
| [MO88] | WR | Exact | None | Yes | Single | $O(n)$ | SP |
| [Ca90] | WR | Exact | None | Yes | Single | $O(n)$ | SP |
| [LM88] | WR | Exact | None | Yes | Single | $O(n)$ | SP |
| [LM88] | WHR | Exact | Not exact | Yes | Single | ? | SP |

(b)

1: Only if some reordering of the circuit has a d-euler trail.
2: Restricted to series-parallel circuits with three or fewer logic levels.
3: Only if the layout height is limited to some small number.

**Fig. 2.10.** Summary of prior functional cell layout methods: (a) static and (b) dynamic designs.

| Layout problem | Width minimization | Height minimization | Circuit reordering | Number of cells | Previously solved for all practical-sized SP circuits |
|---|---|---|---|---|---|
| W | Exact | No | No | Single | Yes |
| WR | Exact | No | Yes | Single | Incompletely (static case) Yes (dynamic case) |
| WH | Exact | Exact | No | Single | No |
| WHR | Exact | Exact | Yes | Single | Incompletely |
| WM | Exact | No | No | Multiple | Yes |
| WRM | Exact | No | Yes | Multiple | No |
| WHM | Exact | Exact | No | Multiple | Incompletely |
| WHRM | Exact | Exact | Yes | Multiple | No |

**Fig. 2.11.** Classification of area-minimization problems and their status.

The solution status of these problems, solved, partially solved, or yet unsolved by prior methods, is given in Fig. 2.11.

The prior methods listed in Fig. 2.10 that claim optimality address five of the eight problem classes. The W problem, namely, exact width minimization for fixed-circuit single cells is solved by two methods [NB85, HS88] for static circuits and by one method for dynamic circuits [Ch90]. A second group of methods addresses the WR problem; three methods [MO88, LM88, Ca90] solve this problem completely for dynamic cells, and one method [NB85] for a subset of all practical-sized static CMOS circuits. A third group partially solves the WHR problem; these methods appear to find exact width and height for a subset of all single-cell practical-sized reordered static CMOS circuits [LC89], [Ma89]. The methods of [LC89, Ma89] are limited to 18% and 56% of all practical-sized circuits, respectively. One method solves the WM problem, achieving exact width minimization for fixed-circuit multiple cells [HH89]. Another method partially solves the WHM problem, finding exact width and height for multiple cells, where the cell height is fixed at a small number [Ba89].

As summarized in Fig. 2.11, the W and WM problems are solved, the WR, WHR and WHM problems are partially solved, and the WH, WRM and WHRM problems have not been addressed by any prior method. It should be noted that a solution to a given problem also solves all its subproblems; for example, a solution to the WRM problem also solves the W, WR and WM problems, since they are restricted cases of it. This book presents exact algorithms that solve all eight problems for all practical-sized dual series-parallel circuits, and the W problem for all practical-sized dual planar circuits.

**Misuse of Term "Optimal."**   In the research literature on the subject of functional cell layout, as in the CAD literature generally, the term "optimal" has often been used with different and misleading meanings. In their seminal paper [Uv81], whose title is "Optimal layout of CMOS functional arrays," Uehara and vanCleemput use the term to describe a layout method that has the potential to produce an optimal solution; when it does find such a solution, the method is able to recognize it as optimal. Those who have not read the paper completely may conclude that it presents an exact algorithm for area minimization. In fact, the authors emphatically state late in the paper that "this heuristic algorithm does not necessarily give the optimal layout. However, if the resulting sequence has no separation areas [i.e., diffusion gaps], it is the real optimal solution." Following Uehara and vanCleemput, several researchers have adopted this looser use of the term optimal. One example states that "in 1981, Uehara and vanCleemput presented an algorithm to *optimally* place CMOS functional cells" (emphasis ours) [He87]. Another example is [KK88], whose title is "A fast heuristic for optimal CMOS functional cell layout generation." The paper itself only claims that its heuristic finds "a *nearly* minimum number of Euler paths" (emphasis ours).  In similar fashion, another paper states that Uehara and vanCleemput "concentrated on ... finding an *optimal* solution" (emphasis ours) [Ma89]. The title of this paper is "A new approach to optimal cell synthesis," but the method presented seems to find an optimal solution only under certain conditions.  Another paper describes its method as "a new methodology which allows the binding of transistor connections to be delayed until an *optimal gate ordering* is determined," (emphasis ours) [CH88] but, in fact, the method is a heuristic.  Other researchers in this field have maintained the traditional, more rigorous, use of the term optimal in reference to their layout methods, using it to denote that their methods *always* finds an optimal solution [LM88], [OL89].

These different uses of the term "optimal" can leave readers wondering whether exact algorithms have been found or not. Unless one reads the literature very carefully, one can easily get the erroneous impression that various layout problems have indeed been solved exactly. When almost every paper uses the term "optimal" in reference to its method, it is no wonder that true optimality is less than fully appreciated!

**Overemphasis on Polynomial Methods.**   Another trend in CAD that discourages the pursuit of exact algorithms is overemphasis on the importance of polynomial-time complexity with the concomitant aversion to any nonpolynomial-time algorithm. This, in part, reflects the tendency of algorithm theorists to describe problems of nonpolynomial-time complexity as "intractable" [GJ79]. For instance, Uehara and vanCleemput propose an exact solution to the functional cell reordering problem, comment on its high complexity, and consequently use it as a justification for their linear-time heuristic [Uv81]. Similarly, other researchers state in reference to this problem, "the search for an arrangement of the transistor pairs that minimizes the area (or equivalently minimizes the diffusion gaps) is a difficult problem, and *very likely intractable*. For this reason, Uehara and vanCleemput introduced a simple heuristic algorithm to align the gates" [MD88] (emphasis ours). They likewise propose a presumably polynomial-time layout heuristic. Another paper, based on a worst-case analysis, states that the exact solution to the reordering problem just discussed is of exponential-time complexity $O(n^n)$ [KK88]; consequently a polynomial-time heuristic is developed to solve it.

The impression is given that polynomial solutions are equated with efficient ones, and nonpolynomial solutions with inefficient ones. However, for any particular problem domain, depending on the limits on the problem size, and the degree and coefficients of the polynomial, a given exponential algorithm can be better or worse than a polynomial-time algorithm. Therefore, the tractability of an algorithm depends both on its time complexity and the size of the problems to which it is applied.

This pessimism is also found in design-related textbooks. One such book asserts that " In general, ... the combinatorial problems involved (are) too time consuming for us to expect to find optimal designs very often. Thus we are usually content with algorithms that produce a 'good' even if not optimal, solution..." [Ul84]. Of course, in general, we agree. It also postulates that "almost all the optimization problems we encounter in VLSI design, as elsewhere, are NP-complete or worse." (It is strongly believed by algorithm theorists that an NP-complete problem cannot be solved by an algorithm with polynomial worst-case time complexity [GJ79].) This sort of pessimistic statement repeated with regularity can discourage researchers from looking for exact algorithms to VLSI problems. If one is taught that something rarely if ever exists, it is not likely one will be looking for it with any hope of success. However, we will demonstrate that our CMOS cell layout algorithms, despite their nonpolynomial-time complexity, are indeed computationally feasible for practical-sized layout problems.

# CHAPTER III

# SERIES-PARALLEL CELL WIDTH

# MINIMIZATION

In 1978, Uehara and vanCleemput published their original paper introducing a new and area-efficient layout style, the one-dimensional functional cell, which we have discussed in Chapters 1 and 2 [Uv78]; Fig. 3.1 illustrates such an cell. They also defined the WR layout problem of Fig. 2.4 and presented a heuristic that attempts to solve it for practical-sized functional cells. Many other researchers have subsequently studied this problem, as we have seen. Some have proposed heuristic solutions to the WR problem, while others have developed algorithms and heuristics for related problems. But no prior methods solve the WR problem exactly as originally proposed.

In this chapter, we present a general and complete algorithmic approach to the W and WR problems for all practical-sized dual series-parallel circuits [MH90]. A new and efficient methodology is presented for finding the minimum number of dual trails that cover a series-parallel multigraph representation $M$ of a transistor circuit and its dual. Two exact algorithms that derive minimum-width layouts from $M$ are presented: *TrailTrace*, which assumes a fixed ordering of $M$, and *R-TrailTrace*, which finds an optimal reordering of $M$. We demonstrate that these algorithms are exact and can handle all circuits of practical size. Results of an exhaustive study of these circuits using programmed implementations of *TrailTrace* and *R-TrailTrace* are presented, which show the importance of reordering for width minimization, and also allow us to compare the effectiveness of prior heuristic approaches to our optimal algorithms.

The chapter is organized as follows: Section 3.1 defines graph optimization problems that model the cell layout problems of interest. Section 3.2 develops the theory of dual trail covering, which is the basis for the algorithms presented in

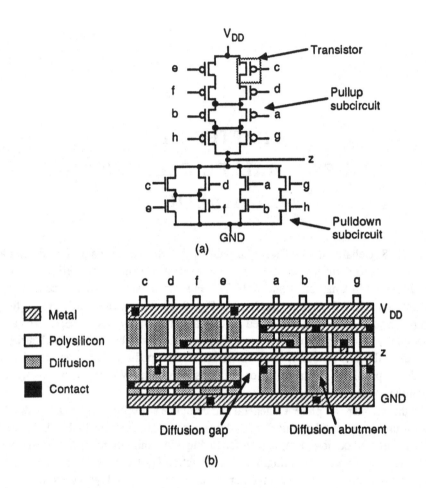

**Fig. 3.1.** CMOS functional cell: (a) transistor circuit; (b) layout in the style of Uehara and vanCleemput.

Sections 3.3 and 3.4 that solve the layout optimization problems exactly. Section 3.5 evaluates previous methods based on the optimal layout data for all practical circuits obtained by our algorithms; it also analyzes the efficiency of *TrailTrace* on these circuits. Section 3.6 extends our approach to minimum-width rows of functional cells.

## 3.1 GRAPH OPTIMIZATION PROBLEMS

We adopt the layout style introduced by Uehara and vanCleemput for the functional cells under consideration. It is based on the assumptions listed in Fig.

2.1, which are summarized here. Static dual series-parallel CMOS circuits with height four or less are addressed, such as the one in Fig. 3.1(a). The pMOS and nMOS transistors in the pullup and pulldown networks are assigned to two separate rows in the layout, so that dual pairs of transistors share the same vertical polysilicon column as their gate terminal; see Fig. 3.1(b). Drain and source terminals connected in the transistor circuit are connected by diffusion abutment if placed adjacent to each other in the cell, or by a horizontal metal wire otherwise; adjacent terminals not connected in the circuit are separated by a diffusion gap, as shown in the figure.

Following [Uv81], the height of the functional cell is assumed to be fixed so that the cell's area depends only on its width. (We will lift this restriction in Chapter 5.) The cell's width $W$ is given by the following formula:

$$W = T + G + 1$$

where $T$ is the number of dual transistor pairs in the pullup and pulldown networks, and $G$ is the number of diffusion gaps required in the layout. The width is in terms of the basic grid size, which is the minimum horizontal spacing between adjacent vertical polysilicon gate lines. A diffusion gap requires the separation between neighboring vertical polysilicon lines to be twice as large as that required by diffusion abutment; see Fig. 3.1(b). Therefore, in the interests of reducing cell area, the preferred method of connecting physically adjacent transistors is diffusion abutment. Thus the main area minimization problem to be solved is finding a permutation of the polysilicon gate columns and an orientation of the transistors that maximizes the number of transistors that can be connected by abutment.

We model the transistor circuit as a graph, and formulate the above area-minimization problem in terms of two related graph-optimization problems. Following [Uv81], labeled undirected *two-terminal series-parallel multigraphs* (TTSPMs) are used, with an edge representing the drain-source diffusion of a transistor and a vertex representing the drain-source interconnections between transistors. (The graph terminology will be that of Harary [Ha69].) An edge is labeled by the name of the input signal on the gate terminal of the transistor it represents. The graph model of the CMOS cells under consideration therefore consists of two labeled dual TTSPMs, $M$ for the nMOS or pulldown part, and $M^d$ for the pMOS or pullup part. The N (north) and S (south) terminals of the pullup graph $M^d$ represent power ($V_{DD}$) and the output $z$, and the N and S terminals of the pulldown graph $M$ denote the output $z$ and ground (GND), respectively. Figures 3.2(b) and 3.2(c) show the multigraphs corresponding to the pMOS and nMOS parts, respectively, of the circuit of Fig. 3.2(a).

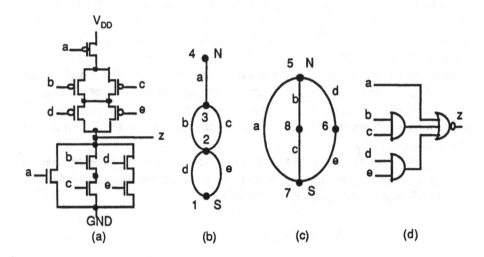

**Fig. 3.2.** Graph model of a cell: (a) transistor circuit; (b) pMOS pullup graph; (c) nMOS pulldown graph; (d) equivalent gate-level logic circuit.

The properties of TTSPMs have been extensively studied in classical switching theory [Ha65]. These graphs are composed from single edges by connecting them together at their terminals through successive applications of the series or parallel composition operations. Dual TTSPMs are composed similarly, except that whenever a series (parallel) operation is performed on two subgraphs, a parallel (series) operation is performed on their duals. As the series and parallel operations implement the AND and OR functions, respectively, we will use the notation * for series and + for parallel composition. We also use the following notation for the simultaneous application of such dual composition operators to $M$ and $M^d$: the *series/parallel operator* */+ represents the simultaneous application of * to $M$ and + to $M^d$; while the *parallel/series operator* +/* represents the simultaneous application of + to $M$ and * to $M^d$. All graphs considered in this chapter are TTSPMs to which these operators always apply.

The structure of a dual series-parallel graph can be represented by a *composition tree T* [LM88]. $T$ is an ordered rooted labeled tree in which a leaf node is labeled with the name of an edge of $M$, and an interior node is labeled with either the */+ or +/* operator applied to the subgraphs represented by its children. The left-right ordering of the children is the order of application of the operator to the children. Figure 3.3(a) shows the composition tree $T$ that models the cell of Fig. 3.2. Changing the order of the children of any node results in a tree $T'$ representing a *reordering M'* of $M$. The tree of Fig. 3.3(b), for example, is a simple reordering of the tree of Fig. 3.3(a). Note that the composition tree has the same structure as a logic-level circuit that is functionally equivalent to $M$ or $M^d$. This relationship is exploited in [Uv81]. This correspondence of structure is illustrated by the logic-level circuit of Fig. 3.2(d) and the composition tree of Fig. 3.3(a). The number of *levels of logic* of a transistor circuit is the depth, or maximum number of nodes from the root to the leaves, of its composition tree. For example, the circuit of Fig. 3.2(a) has three levels of logic because its composition tree in Fig. 3.3(a) has a depth of three.

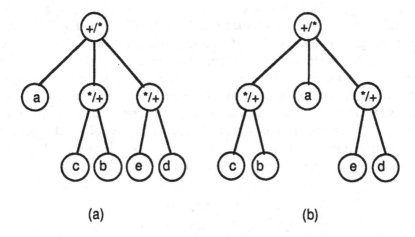

(a)                              (b)

**Fig. 3.3.** (a) Composition tree $T$ for Fig. 3.2. (b) Composition tree $T'$ after reordering.

An equivalent representation of the structure of $M$ is a *composition expression* $E$, which is simply the preordering of the node labels of $T$ [MH87]. The composition expression corresponding to Fig. 3.3(a) is, in infix notation,

$$((a) +/* (c */+ b) +/* (e */+ b))$$

and, in the slightly more concise prefix notation,

$$(+/* (a)(*/+ cb)(*/+ ed))$$

The reordered prefix expression corresponding to Fig. 3.3(b) is

$$(+/* (*/+ cb)(a)(*/+ ed))$$

which reflects a simple permutation of the three operands of the +/* operator, forming the root node of the composition tree.

The order of placement of transistors in the two horizontal rows of the functional cell layout affects the cell's width, because it determines which of the physically adjacent transistors can be connected by abutment. Considering only a single graph for a moment, all the transistors whose edges compose a trail in the circuit graph can be connected by abutment if their order in the trail sequence is the same as their order in the layout. Due to the vertical alignment restriction (assumption 6 of Fig. 2.1), the sequence of edges and vertices must define dual trails in both the pMOS and nMOS multigraph. In Fig. 3.2, for example, the trails $4a3b2c3$ in Fig. 3.2(b) and $7a5b8c7$ in Fig. 3.2(c) are dual trails; for brevity, we refer of them together as the *dual trail* $t_1 = abc$. In the cell layout, the transistors corresponding to a dual trail are connected by diffusion abutment. Neighboring groups of transistors corresponding to disconnected dual trails are separated by a diffusion gap; see Fig. 3.1(b). A set of dual trails is said to *cover M* (and hence $M^d$) if it contains all the edges of $M$ exactly once. Therefore, the problem of minimizing the area of a cell under the above assumptions reduces to finding the minimum number of dual trails that cover the transistor multigraphs representing the pullup and pulldown subcircuits.

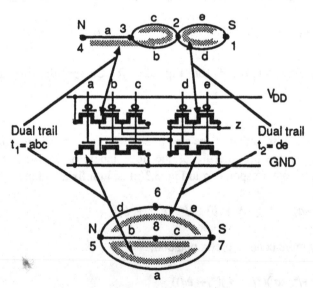

**Fig. 3.4.** Correspondence between dual trails and transistor layout.

The number of trails required to cover a given transistor circuit depends on the ordering and orientation of the series-parallel subgraphs in the multigraph model. The circuit of Fig. 3.2, for example, requires two dual trails $t_1 = abc$ and $t_2 = de$ to cover the dual multigraphs of Figs. 3.2(b) and 3.2(c). Figure 3.4 shows the correspondence between dual trails in the multigraph model and the transistor rows in the cell layout. It can readily be shown that by swapping the subcircuits *a* and *bc* in Fig. 3.2(a), as illustrated in Fig. 3.3, the multigraph model of the reordered circuit can be covered by one dual trail *t* = *bcade*. Like [CH88, MD88, NB85, Uv81], we also address the problem of finding a reordering *M'* that minimizes the number of dual trails needed to cover *M*, which corresponds to the WR layout problem of Fig. 2.4. We define this as a graph optimization problem and present the algorithm *R-TrailTrace* that solves it exactly.

In some instances, designers choose a specific transistor ordering that does not have the absolute minimum number of dual trails. For example, they fix the transistor order to minimize propagation delay through the cell. Therefore, we also define a second graph optimization problem of finding the minimum number of dual trails that cover a given *M*, which corresponds to the W layout problem of Fig. 2.4. We also present the algorithm *TrailTrace* that solves this problem optimally.

## 3.2 THEORY OF DUAL TRAIL COVERING

We now develop a general theory of dual trail covering that identifies all equivalence classes of trail covers, and precisely quantifies the relationship between our method and that of Uehara and vanCleemput [Uv81]. These equivalence classes also lead to efficient algorithms that solve the W and WR layout problems. The basic idea is that only one trail cover from each equivalence class of covers of a graph is required to represent all possible optimal layouts for that graph. As the graph is composed in bottom-up dynamic programming fashion, these trail covers are concatenated to form new ones, but their number never exceeds the number of trail-cover partitions.

First, we present some formal definitions and illustrate the main concepts. To facilitate the following discussion concerning dual trails and their endpoints, we represent a *dual trail* by an alternating sequence of vertex and edge pairs $(v_1, v_1^d)$, $(e_1, e_1^d)$, ... ,$(e_{n-1}, e_{n-1}^d)$, $(v_n, v_n^d)$, where the sequence of first elements in the pairs $(v_1, e_1, \ldots e_{n-1}, v_n)$ is a trail *t* in *M* and the sequence of second elements $(v_1^d, e_1^d, \ldots, e_{n-1}^d, v_n^d)$ is a trail $t^d$ in $M^d$, and $e_i = e_i^d$, for $1 \leq i \leq n-1$. For example, a dual trail in Fig. 3.2 is $t_1$ = (4,7),(a,a),(3,5),(b,b),(2,8), (c,c), (3,7). Where the dual trail in question is clear, we will denote it concisely by its edge sequence only, as in $t_1$ = *abc* for the preceding instance. A dual trail of *M* with at least one pair of endpoints composed of N or S terminals of *M* is called a *distinguished trail*, otherwise it is called an *internal trail*. A *dual-euler* or *d-euler trail* [NB85] is any dual trail in $M$ and its dual $M^d$ that contains all their edges exactly once.

Consider the concatenation of dual trails in different graphs when the graphs are combined using the connection operators */+ and +/*. We begin by considering the trails in $M$ apart from those in its dual. Let $t_1$ and $t_2$ be trails in subgraphs $M_1$ and $M_2$, respectively, of $M$, and let $M_3 = M_1 \# M_2$, where # denotes * or +. Trails $t_1$ and $t_2$ can be concatenated to form a longer trail $t_3$ in $M_3$, if each of the trails has as an endpoint a terminal of $M_1$ and $M_2$, respectively, that is connected in $M_3$ with the terminal of the other trail. In this case, $t_1$ and $t_2$ are called *compatible* trails, which implies that they can be concatenated to form a single trail. The concatenation of dual trails $t_1$ and $t_2$ is similar, except that the above condition on the endpoints must hold in the duals also; $t_1$ and $t_2$ are then *compatible dual trails*. A *trail cover TC* of a dual TTSPM $M$ is a set of pairwise incompatible dual trails $\{t_1, t_2, ..., t_k\}$ for which there is a one-to-one correspondence between the edges in $M$ and the edges of the trails in the trail cover. For example, consider Fig. 3.5, which illustrates the operation $M_3 = M_1$ */+ $M_2$. Here $M_1$ has trail cover $\{t_1\}$ and $M_2$ has trail cover $\{t_2\}$.

$$\{t_1\} = \{(2,4),(a,a),(1,5),(b,b),(2,6)\}$$
$$\{t_2\} = \{(2,6),(c,c),(3,4)\}$$

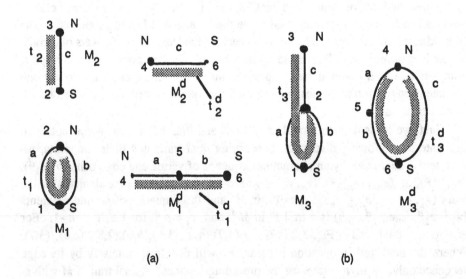

(a)                                                                (b)

**Fig. 3.5.** Series-parallel connection of two graphs and their dual trails.

A trail cover of $M_3$ is

$$\{t_3\} = \{(2,4),(a,a),(1,5),(b,b),(2,6),(c,c),(3,4)\}$$

since $t_1$ and $t_2$ are compatible. But, given that the trail covers for $M_1$ and $M_2$ are

$$\{t'_1\} = \{(1,4),(a,a),(2,5),(b,b),(1,6)\}$$
$$\{t'_2\} = \{(2,6),(c,c),(3,4)\}$$

respectively, then the resulting trail cover for $M_3$ is

$$\{t'_1, t'_2\} = \{(1,4),(a,a),(2,5),(b,b),(1,6); \quad (2,6),(c,c),(3,4)\}$$

which is a set of two separate dual trails, since these trails are not compatible.

As illustrated in the preceding example, a graph $M$ can have multiple trail covers. For the purposes of dual trail concatenation, all trail covers of $M$ that have trails with exactly the same pairs of terminals can be considered equivalent. In order to minimize the set of trail covers describing $M$ without eliminating any that may eventually lead to an optimal solution to the layout problem, we partition the set of trail covers into such equivalence classes. Only a trail cover with the smallest number of trails in each equivalence class need be retained as the representative of the class, and all others can be safely discarded. This is the key to the efficiency of the algorithms *TrailTrace* and *R-TrailTrace* presented below. Before we define these equivalence classes formally, however, we must first introduce the concept of trail type.

The ability of a dual trail $t$ in $M$ to concatenate with dual trails of another multigraph that is connected to $M$ using */+ or +/*, is determined by the endpoints of $t$. There are three *vertex types* in $M$, namely N, S and I, where N (north) and S (south) are the usual terminals of $M$, and I (internal) denotes any nonterminal vertex of $M$. Let $t$ be a dual trail in $M$. An ordered pair $(A_i, B_i)$ of vertex types defines the *endpoint type* of a pair of endpoints $(v_i, v_i^d)$ of a dual trail with respect to its ability to concatenate. Such a pair is a member of the set $\{(N,N), (N,S), (S,N), (S,S), (I,I)\}$. The first four pairs characterize an end of a dual trail that can concatenate with other trails. All other pairs of endpoints are placed in the (I,I) class, which indicates that at least one of the two endpoints of a trail $t$ in either of the dual graphs forming $M$ is internal (I), and so is not accessible for concatenation with dual trails of other multigraphs. Hence (I,I) represents each of the endpoint pairs (N,I), (S,I), (I,N), (I,S).

Let $t_1 = (v_1, v_1^d), (e_1, e_1^d), \dots, (e_{n-1}, e_{n-1}^d), (v_n, v_n^d)$ be a dual trail. Then the *trail type* $T(t_1)$ of $t_1$ is the unordered pair of endpoint types, denoted

$(A_1,B_1)/(A_n,B_n)$, where $(A_1,B_1)$ and $(A_n,B_n)$ are the types of the endpoint pairs $(v_1,v_1^d)$ and $(v_n, v_n^d)$, respectively. As observed above, each endpoint type can assume five possible values, namely: (N,N), (N,S), (S,N), (S,S), (I,I). Since the order of the endpoints in a trail type is irrelevant, types such as (N,N)/(I,I) and (I,I)/(N,N) are equivalent. Pairs of distinct endpoint types may therefore be combined to form ten nonequivalent trail types, which we denote as follows: (N,N)/(N,S), (N,N)/(S,N), (N,N)/(S,S), (N,N)/(I,I), (N,S)/(S,N), (N,S)/(S,S), (N,S)/(I,I), (S,N)/(S,S), (S,N)/(I,I), (S,S)/(I,I). Only (I,I) may be paired with itself to form an additional *internal trail type* (I,I)/(I,I), leading to the conclusion that only the foregoing 11 different trail types can occur in dual multigraphs. We also use Z to denote all trail covers that contain two or more trail types, as will be explained later.

The completeness of the above set of trail types follows from the duality of the subgraphs forming $M$. (N,N)/(S,S) and (N,S)/(S,N) are the types that characterize the dual trails of a single-edge graph and its dual. The */+ and +/* operators can be applied to these two trail types successively to compute the closed set of trail types listed above for all dual TTSPMs. For example, for the trails shown in Fig. 3.5, we have

$$
\begin{aligned}
T(t_3) &= T(t_1) \text{ */+ } T(t_2) \\
&= (N,N)/(N,S) \text{ */+ } (N,N)/(S,S) \\
&= (N,N)/(I,I)
\end{aligned}
$$

since the (N,S) endpoint type of $t_1$ concatenates with (S,S) of $t_2$. Only an (N,X) endpoint type of the left operand of the */+ operator and an (S,X) type of the right operand can be concatenated, where X represents either N or S. The resulting trail type naturally has two endpoint types, the two that were not concatenated. If concatenation is not possible, then the result is Z, which represents all trail covers having multiple trails. For example, the trail-type concatenation of trails in Fig. 3.4 is

$$
\begin{aligned}
T(t_2,t_1) &= T(t_2) \text{ */+ } T(t_1) \\
&= (S,N)/(S,S) \text{ */+ } (N,S)/(I,I) \\
&= Z
\end{aligned}
$$

since $t_1$ and $t_2$ do not concatenate. Figure 3.6 shows a complete operation table for the */+ operator acting on the 11 trail types and Z.

We can now partition trail covers according to the trail types of their incompatible distinguished dual trails. Let $tc = \{dt_1, dt_2, ... , dt_j, it_1, it_2, ... , it_k\}$, where $dt_n$ is a distinguished dual trail, and $it_n$ is an internal trail of type (I,I)/(I,I). The *trail-cover type* $T(tc)$ of $tc$ is defined to be $\{T(dt_1), T(dt_2), ... ,T(dt_j)\}$. Note that

| */+ | (N,N)/(S,S) | (N,S)/(S,N) | (N,N)/(S,N) | (N,S)/(S,S) | (N,N)/(N,S) | (S,N)/(N,S) | (N,S)/(L,L) | (N,N)/(L,L) | (S,S)/(L,L) | (L,L)/(L,L) | Z |
|---|---|---|---|---|---|---|---|---|---|---|---|
| (N,N)/(S,S) | N | (N,N)/(S,S) | N | (S,S)/(L,L) | N | (S,S)/(L,L) | N | N | N | N | N |
| (N,S)/(S,N) | (N,N)/(S,S) | (N,S)/(S,S) | (N,S)/(S,N) | N | (S,N)/(L,L) | N | (S,N)/(L,L) | N | N | N | N |
| (N,N)/(S,N) | N | (N,N)/(S,N) | N | (S,N)/(L,L) | N | (S,N)/(L,L) | N | N | N | N | N |
| (N,S)/(S,S) | (N,N)/(S,S) | (N,S)/(S,N) | (N,S)/(S,S) | N | (S,S)/(L,L) | N | (S,S)/(L,L) | N | N | N | N |
| (N,N)/(N,S) | (N,N)/(L,L) | (N,S)/(L,L) | (N,N)/(L,L) | (N,S)/(L,L) | (L,L)/(L,L) | (L,L)/(L,L) | N | N | N | N | N |
| (S,N)/(N,S) | N | N | N | N | N | N | N | N | N | N | N |
| (S,S)/(S,S) | N | (N,S)/(L,L) | (N,N)/(L,L) | N | (L,L)/(L,L) | (L,L)/(L,L) | N | N | N | N | N |
| (N,N)/(L,L) | (N,N)/(L,L) | (N,S)/(L,L) | (N,N)/(L,L) | N | (L,L)/(L,L) | (L,L)/(L,L) | N | N | N | N | N |
| (N,S)/(L,L) | N | N | N | N | N | N | N | N | N | N | N |
| (S,N)/(L,L) | N | N | N | N | N | N | N | N | N | N | N |
| (S,S)/(L,L) | N | N | N | N | N | N | N | N | N | N | N |
| (L,L)/(L,L) | N | N | N | N | N | N | N | N | N | N | N |
| Z | N | N | N | N | N | N | N | N | N | N | N |

**Fig. 3.6.** Operation table for the */+ operator applied to the 11 dual trail types and Z.

$T(tc)$ consists only of the trail types of the distinguished dual trails in $tc$. The internal trail type (I,I)/(I,I) is not included in the trail-cover type because it cannot be concatenated with other trails.

These concepts are illustrated by Fig. 3.7, whose multigraphs can be covered by the two distinguished trails

$$dt_1 = (1,6), (c,c), (2,7), (d,d), (1,9)$$
$$dt_2 = (5,6), (e,e), (2,8), (f,f), (5,9)$$

and the internal trail

$$it_1 = (1,10), (a,a), (3,9), (b,b), (5,10), (h,h), (4,11), (g,g), (1,10)$$

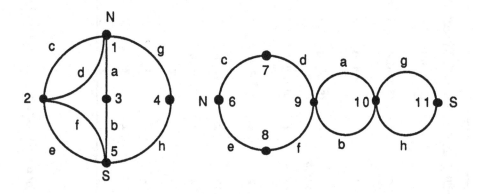

**Fig. 3.7.** Graph model for the transistor circuit of Fig. 3.1(a).

Clearly, $T(dt_1) = $ (N,N)/(I,I), $T(dt_2) = $ (S,N)/(I,I), and $T(it_1) = $ (I,I)/(I,I). If $tc = \{dt_1, dt_2, it_1\}$, then $T(tc) = \{$(N,N)/(I,I), (S,N)/(I,I)$\}$.

The two key characteristics of a given trail cover $tc$ of a subgraph of $M$ are its trail-cover type $T(tc)$ and its number of internal trails. All trail covers that are the same in these two respects are equivalent as far as finding an optimal trail cover of $M$ is concerned. Furthermore, any trail cover with a nonminimum internal trail count for its type cannot possibly contribute to finding the optimal trail cover of $M$. Therefore, only a representative optimal trail cover from each class of equivalent types needs to be retained at each step in the graph composition process to guarantee that the optimal trail cover of $M$ will be found.

1.  (N,N)/(S,S)
2.  (N,S)/(S,N)
3.  (N,N)/(S,N)
4.  (N,S)/(S,S)
5.  (N,N)/(N,S)
6.  (S,N)/(S,S)
7.  (N,N)/(I,I)
8.  (N,S)/(I,I)
9.  (S,N)/(I,I)
10. (S,S)/(I,I)
11. (I,I)/(I,I)
12. (N,N)/(N,S);          (S,N)/(S,S)
13. (N,N)/(S,N);          (N,S)/(S,S)
14. (N,N)/(S,S);          (N,S)/(I,I)
15. (N,N)/(S,S);          (S,N)/(I,I)
16. (N,S)/(S,N);          (N,N)/(I,I)
17. (N,S)/(S,N);          (S,S)/(I,I)
18. (N,N)/(N,S);          (S,N)/(I,I)
19. (N,N)/(N,S);          (S,S)/(I,I)
20. (S,N)/(S,S);          (N,N)/(I,I)
21. (S,N)/(S,S);          (N,S)/(I,I)
22. (N,N)/(S,N);          (N,S)/(I,I)
23. (N,N)/(S,N);          (S,S)/(I,I)
24. (N,S)/(S,S);          (N,N)/(I,I)
25. (N,S)/(S,S);          (S,N)/(I,I)
26. (N,N)/(I,I);          (N,S)/(I,I)
27. (N,N)/(I,I);          (S,N)/(I,I)
28. (N,N)/(I,I);          (S,S)/(I,I)
29. (N,S)/(I,I);          (S,N)/(I,I)
30. (N,S)/(I,I);          (S,S)/(I,I)
31. (S,N)/(I,I);          (S,S)/(I,I)
32. (N,S)/(S,N);          (N,N)/(I,I);          (S,S)/(I,I)
33. (N,N)/(S,S);          (N,S)/(I,I);          (S,N)/(I,I)
34. (N,N)/(N,S);          (S,N)/(I,I);          (S,S)/(I,I)
35. (S,N)/(S,S);          (N,N)/(I,I);          (N,S)/(I,I)
36. (N,N)/(S,N);          (N,S)/(I,I);          (S,S)/(I,I)
37. (N,S)/(S,S);          (N,N)/(I,I);          (S,N)/(I,I)
38. (N,N)/(I,I);          (N,S)/(I,I);          (S,N)/(I,I)
39. (N,N)/(I,I);          (N,S)/(I,I);          (S,S)/(I,I)
40. (N,N)/(I,I);          (S,N)/(I,I);          (S,S)/(I,I)
41. (N,S)/(I,I);          (S,N)/(I,I);          (S,S)/(I,I)
42. (N,N)/(I,I);          (N,S)/(I,I);          (S,N)/(I,I);          (S,S)/(I,I)

**Fig. 3.8.** The 42 dual trail-cover types.

The operators +/* and */+ can be applied to two cover-type operands to compute the cover type resulting from the concatenation of the trails that constitute the covers. We illustrate this for the series/parallel operator */+. Let $T(tc_1) = \{(N,N)/(N,S), (S,N)/(I,I)\}$ and $T(tc_2) = \{(S,N)/(I,I)\}$. Then

$$
\begin{aligned}
T(tc_1) \; */+ \; T(tc_2) &= \{(N,N)/(N,S), (S,N)/(I,I)\} \; */+ \; \{(S,N)/(I,I)\} \\
&= \{(N,S)/(I,I), (S,N)/(I,I)\}
\end{aligned}
$$

since the (N,N) terminals in $T(tc_1)$ connect to the (S,N) terminals in $T(tc_2)$ when their corresponding graphs are combined by the */+ operator. Applying the */+ and +/* operations successively to the set of trail-cover types until closure is achieved, leads to the following theorem.

**Theorem 3.1:** There are exactly 42 dual trail-cover types.

The set of 42 dual trail-cover types constitutes a pair of semigroup algebras with respect to the */+ and +/* operators. Eleven of the 42 types contain one trail type each; these are, of course, the 11 dual trail types derived above. The remaining 31 contain 2, 3, or 4 trail types; see Fig. 3.8 for a complete listing. These 31 multiple-trail-cover types are a refinement of Z in Fig. 3.6. The 42 trail-cover types define equivalence classes on the trail covers with respect to concatenation.

A *complete trail cover CTC* of $M$ is a set of trail covers of $M$ that contains a representative of each possible cover type of $M$. A *minimum* CTC of $M$ is a CTC in which each of the trail covers has the minimum number of internal trails with respect to the trail covers of each type in $M$. A *unary* minimum CTC (UMCTC) of $M$ is a minimum CTC that contains only one representative cover of each possible cover type.

The above trail-cover analysis allows us to pinpoint the limitations of the approach of Uehara and vanCleemput. They observed (Theorem 2 of [Uv81]) that if every internal node of the composition tree of $M$ has an odd number of children, then $M$ has a d-euler trail. If a given graph does not have this type of composition tree, the Uehara-vanCleemput heuristic transforms it into a "pseudograph" having the above property by adding "pseudoedges", and systematically traces a single d-euler trail through the pseudograph. In our notation, $M$ has the following CTC consisting of just two d-euler trail types:

$$
ctc_0 = \{(N,N)/(S,S), (N,S)/(S,N)\}
$$

As we have seen, to characterize all possible TTSPMs with no restrictions on their composition trees requires the 42 trail-cover types identified in Fig. 3.8, of which $ctc_0$ is but a very small subset. Since our approach is based on this complete set of trail-cover types, it can find an optimal trail cover in all situations, including those that cannot be handled by the Uehara-vanCleemput and other heuristic methods.

## 3.3  OPTIMAL TRAIL COVERING WITHOUT REORDERING

We now describe an algorithm *TrailTrace* based on the foregoing theory that efficiently computes a UMCTC for the multigraph $M$ in systematic, bottom-up fashion. The algorithm, which is summarized in Fig. 3.9, begins by constructing the composition tree $T$ for $M$. It then computes the UMCTC for $M$ by visiting the nodes of $T$ in postorder using the function Node_CTC recursively. In visiting the nodes, *TrailTrace* either assigns the primitive UMCTC to each of the single-edge subgraphs at the leaves of $T$ or calls the function Ordered_children_CTC on the internal nodes of $T$. Ordered_children_CTC computes the UMCTC for a node from those of its children subgraphs in left-to-right order. This function processes these children, concatenating each trail cover of a child with each trail cover of the UMCTC representing all the children to its left. It retains the trail covers with minimum internal trail count from each new trail-cover type generated. The final step is to choose a trail cover of $M$ that has the fewest dual trails from the UMCTC of M. This leads to the following theorem.

**Theorem 3.2.** The algorithm *TrailTrace* computes a minimum dual trail cover for any $M$. Therefore, it solves the W layout minimization problem exactly for any dual series-parallel circuit.

*TrailTrace* has time complexity $O(n)$, where $n$ is the number of transistors in the cell under consideration. This follows from the fact that the composition tree $T$ of a multigraph of $n$ edges has $n$ leaves corresponding to transistors. The UMCTC associated with any leaf or internal node can contain no more than 42 different types, since only the trail cover of a given type with the fewest internal trails is retained. Each of the $n - 1$ concatenation operations represented by */+ and +/* in $T$ requires constant time, since at most $42 \times 42 = 1764$ separate trail-cover concatenations need be performed. Therefore, in the worst case, the time complexity of *TrailTrace* is $1764(n - 1) = O(n)$. In practice, we have observed that the constant is far less than 1764.

---

```
function  Ordered_ children_CTC(Node_type, Child_CTC, Child_order):  CTC_type;
begin
Ordered_ children_CTC := Child_CTC[Child_order[1]];
for  i := 2 to  Child_count  of Node  do
     begin
     Temp_ children_CTC := empty;
     for each trail cover in Ordered_children_CTC do
          for each trail cover in Child_CTC[Child_order[i]] do
               begin
               Concatenate both trail covers according to Node_type operation;
               if resulting TC type does not exist in Children_CTC then
                    Add trail cover to Temp_children_CTC
               else if resulting trail cover has fewer internal trails than does TC
               with same TC type in Temp_ children_CTC  then
                    Substitute resulting  cover for current cover in Temp_children_CTC;
               end;
          Ordered_ children_CTC := Temp_ children_CTC;
          end;
end; {Ordered_ children_CTC}

function  Node_CTC(Node):  CTC_type;
begin
Node_CTC := empty;
if Node is a leaf  then
     Node_CTC := Edge_CTC   {Edge_CTC consists of the two covers of an edge}
else
     begin
     for  i := 1 to  Child_count  of Node  do
          Child_CTC[i] := Node_CTC(Child[i]);
2:   for each distinct Child_order do    {This statement is not in TrailTrace}
          begin
          Temp_node_CTC:=Ordered_children_CTC(Node_type,Child_CTC,
               Child_order);
3:        Update Node_CTC with new or better trail covers from Temp_node_CTC;
          end;
     end;
end; {Node_CTC}

procedure  (R-)TrailTrace(M);
begin
Construct composition tree T from M;
Graph_CTC := Node_CTC(T);
1: Choose for layout the smallest trail cover from Graph_CTC;
   end.  {(R-)TrailTrace}
```

---

**Fig. 3.9.** *TrailTrace* and *R-TrailTrace* layout algorithms. (Only *R-TrailTrace* has statement 2.)

To illustrate the effectiveness of the trail-covering method, consider the example of Fig. 3.10, which is Fig. 14 of [Uv81]. The Uehara-vanCleemput heuristic method is unable to find the d-euler trail in the dual multigraph pair of Fig. 3.10(b). This is because the method does not handle subgraphs with even numbers of edges in an optimal fashion. It erroneously permutes the components of Fig. 3.10(b) into the reordered form shown in Fig. 3.7. A minimum set of trails for the latter multigraph pair consists of two dual trails: *abhg* and *cdfe*. The heuristic reordering rule of Chen and Hou [CH88] also fails to guarantee the correct reordering of this circuit, because it is based on reducing the number of vertices of odd degree by reordering. In this particular example, as in many others, all the vertices are of even degree in all the reorderings. However, the original arrangement of Fig. 3.10(b) has a d-euler trail *hgdcabef* that is found by *TrailTrace* as follows.

Let $ctc_i$ be the UMCTC for $M_i$ in Fig. 3.10(b).

$$ctc_1 = \{tc_1, tc_2, tc_3, tc_4\}$$
$$ctc_2 = \{tc_5, tc_6\}$$
$$ctc_3 = \{tc_7, tc_8, tc_9, tc_{10}\}$$

Now the trail covers in $ctc_1$ of $M_1$ are

$$tc_1 = \{(1,6), (a,a), (2,7); \quad (5,7), (b,b), (2,6)\}$$
$$tc_2 = \{(1,6), (a,a), (2,7), (b,b), (5,6)\}$$
$$tc_3 = \{(1,7), (a,a), (2,6); \quad (5,6), (b,b), (2,7)\}$$
$$tc_4 = \{(1,7), (a,a), (2,6), (b,b), (5,7)\}$$

The set $ctc_2$ of trail covers for $M_2$ is $\{tc_5, tc_6\}$ where

$$tc_5 = \{(1,10), (d,d), (3,8), (c,c), (1,7); \quad (5,10), (f,f), (3,9), (e,e), (5,7)\}$$
$$tc_6 = \{(3,7), (c,c), (1,8), (d,d), (3,10), (f,f), (5,9), (e,e), (3,7)\}$$

The third set $ctc_3$ is obtained similarly. The computation of the UMCTC for $M$ performed by *TrailTrace* consists of concatenating each combination of trail-covers from $ctc_1$, $ctc_2$ and $ctc_3$ in two steps. In the first step, the trail-covers $tc_5$ and $tc_4$ concatenate to form a single trail $\{(1,10), (d,d), (3,8), (c,c), (1,7), (a,a), (2,6), (b,b), (5,7), (e,e), (3,9), (f,f), (5,10)\}$. When concatenated with $tc_8$ in the final step, this trail leads to the one-member trail cover $\{(5,10), (h,h), (4,11), (g,g), (1,10), (d,d), (3,8), (c,c), (1,7), (a,a), (2,6), (b,b), (5,7), (e,e), (3,9), (f,f), (5,10)\}$, which is of type

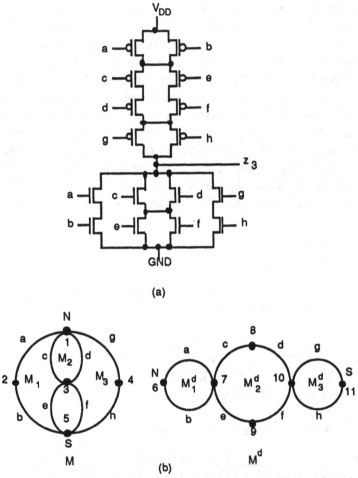

**Fig. 3.10.** Circuit for which prior heuristic methods cannot find an optimal layout.

(I,I)/(I,I). In fact, notice that the beginning and endpoint pairs are identical, making this a d-euler circuit. *TrailTrace* breaks such circuits in an arbitrary spot, since this break has no effect on cell width. Let the circuit be broken between edges *a* and *c*. The corresponding edge-sequence *abefhgdc* is an optimal d-euler trail. The layout of this solution appears in Fig. 3.11. Its area is about 10% less than the Uehara-vanCleemput solution given in Fig. 3.1(b), due to its use of one less diffusion gap. The computation time to find the minimum trail covers using the computer program that implements *TrailTrace* is insignificant (a fraction of a second on a mainframe computer).

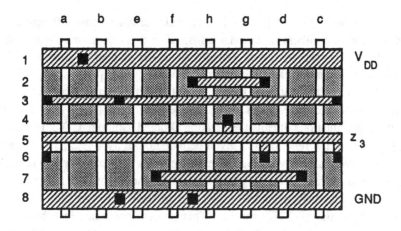

**Fig. 3.11.** Minimum-width layout of the circuit of Fig. 3.10.

## 3.4 OPTIMAL TRAIL COVERING WITH REORDERING

The *TrailTrace* algorithm finds the optimal trail cover for a given $M$, and so solves the W layout problem. If *TrailTrace* is applied to the circuit of Fig. 3.7, it finds a layout having two trails instead of one, whereas it finds a single trail cover when given the reordered graph of Fig. 3.10(b). Our experiments reported below show that in 61% of circuits of height four or less, a reordering $M'$ of $M$ has a smaller optimal trail cover than $M$. The heuristic presented in [Uv81] is guaranteed to find these reorderings only if the graph has the property that every internal node of the composition tree of $M$ has an odd number of children. The reordering algorithm of Nair et al. finds the optimal reordering only if the graph has a d-euler trail in at least one of its reorderings [NB85]. Only our algorithm finds the optimal trail cover for any dual TTSPM under all reorderings $M'$ of $M$.

We now present the algorithm *R-TrailTrace* that finds the optimal trail cover for any reordering. It is based on *TrailTrace* with one important modification. At any interior node in a composition tree $T$, we permute all the isomorphic children and compute the function Ordered_children_CTC for each such reordering, which generates a UMCTC for the subgraph corresponding to that node. This is accomplished by the for loop (statement 2) in the function Node_CTC in Fig. 3.9. This loop is present in *R-TrailTrace* and not in *TrailTrace*. The only salient data saved for a given permutation are the new trail-cover types it produces and the number of internal trails for each type, if better than what has been saved thus far. This is done by the Update Node_CTC step (statement 3) of function Node_CTC.

Different permutations of the children typically produce some different cover types with varying numbers of internal trails. If two permutations produce trail covers of the same type, the trail cover with the greater number of internal trails is discarded, since it cannot yield the optimal solution. Let $S_1, S_2, ..., S_m$ be the children of a node representing a subgraph $S$ of $M$ in a composition tree $T$. *R-TrailTrace* systematically generates all distinct permutations of $S_1, S_2, ..., S_m$. For each of these permutations, it calls the function Ordered_children_CTC. Following each application of the function to a given permutation of $S_1, S_2, ..., S_m$, we save only the trail covers of the types that have not yet been generated for any previous permutation, and we replace existing trail covers with those that have fewer internal trails than any previously generated trail cover of their type. At the end of this process, we have the UMCTC for $S$. This process continues in a bottom-up fashion. The final result is the UMCTC for the best reordering $M'$ of $M$. This leads to the following result.

**Theorem 3.3:** The algorithm *R-TrailTrace* computes the minimum dual trail cover for any reordering $M'$ of $M$. Therefore, it solves the WR layout minimization problem exactly for any dual series-parallel circuit.

Theorem 3.3 follows directly from the fact that *R-TrailTrace* generates all relevant orderings of the children of each internal node of $T$. Note that it does not generate every possible reordering of $M$; the number of such reorderings is prohibitively large even for practical circuits. *R-TrailTrace* has time complexity $O(h!n)$, where $h$ is the maximum height of the circuit, which is the maximum number of transistors along any path from the circuit output to $V_{DD}$ or GND, i.e., between the N and S terminals in the corresponding TTSPM, and $n$ is the number of transistors in the circuit. Although, in the worst case, the number of reorderings increases exponentially with circuit size, the latter is constrained in practice by the height $h$ of the circuit, which, for practical circuits, is typically limited to four. In the next section, we show through extensive computational studies that *R-TrailTrace* efficiently handles all circuits of practical size.

## 3.5 ANALYSIS OF COMPLETE CLASS OF PRACTICAL CELLS

As just noted, the number of distinct circuits of practical size is relatively small. For example, we have counted that there are 3503 functionally different circuits in the class of dual series-parallel CMOS circuits with height $h \leq 4$; also see [De87]. We conducted an extensive series of computer experiments to explore the entire space of such circuits. In addition to implementing *TrailTrace* and *R-TrailTrace*, we have programmed the Uehara and vanCleemput heuristic [Uv81]. We generated each circuit in the foregoing height-limited class, analyzed the layout produced for it by the various methods and compared the run-time of our two algorithms to each other. This analysis took approximately 30 hours on a MicroVAX II minicomputer rated at 0.9 MIPS. The results obtained are discussed below.

Figure 3.12 summarizes the key properties of the 3503 circuits examined. The vast majority of these circuits have a height of four; see Fig. 3.12(a). As Fig. 3.12(b) shows, 56% of the circuits are d-eulerian in some ordering of the circuit, whereas 44% are not. A circuit is considered to be *irreducible* if and only if the minimum number of dual trails in its cover is the same for all reorderings. Figure 3.12(c) shows how many circuits have a given number of levels of logic. This refers to the maximum number of internal nodes along any path from the root node to a leaf node of the composition tree $T$. The levels of logic range from one to six. For example, the circuit of Fig. 3.2 has 3 levels as can be seen immediately from Fig. 3.3. It is significant that the layout method of Lefebvre and Chan [LC89] only handles circuits with three or fewer levels of logic. Therefore, according to our analysis in Fig. 3.12(c), their method is valid for only 18% of the practical-sized functional cells. Figure 3.12(d) divides the circuits into four classes A, B, C and D based on whether they are irreducible and/or d-eulerian for some reordering.

The trail reduction and area improvements found by our experiments are summarized in Fig. 3.13. This figure compares our algorithms to each other, to the heuristics of [Uv81] and [CH88], and to the reordering algorithm of Nair et al. [NB85]. We found that, through reordering, *R-TrailTrace* can reduce the layout of 61% of the 3503 circuits by one or more trails over *TrailTrace*; see Fig. 3.12(a). This corresponds to the circuits of classes C and D in Fig. 3.12(d). The resulting area improvements can be more than 20%, as Fig. 3.13(b) demonstrates. The optimal layout for the remaining 39% of the circuits, all of which are irreducible, can be found by either of our algorithms. These are the circuits forming classes A and B. Note that *R-TrailTrace* is unique in finding the optimal layout for every circuit in each of the four classes A-D.

The Uehara-vanCleemput method misses the optimal layout for 51% of the circuits; see Fig. 3.13(a). In these cases, *R-TrailTrace* produces area improvements from five to more than 20% over the heuristic of [Uv81]; see Fig. 3.13(b). The algorithm of Nair et al. [NB85] yields the optimal layout by reordering if and only if there exists an ordering that results in a d-euler trail cover of the graph model, i.e., a single dual trail covering the entire graph. This property holds for the circuits of classes B and C of Fig. 3.12(d), which account for only 56% of the circuits studied. By reordering, the Nair et al. method can improve the layout of only the 45% of circuits constituting class C. Therefore, like the heuristic of [Uv81], this method is effective on only about half of the practical circuits. The Nair et al. method produces

| Height $h$ of circuit | Number of circuits |
|---|---|
| 1 | 1 |
| 2 | 6 |
| 3 | 80 |
| 4 | 3416 |

(a)

| Minimum number of trails in cover | Percent of circuits |
|---|---|
| 1 | 56 |
| 2 | 39 |
| 3 | 5 |
| 4 | 0.2 |

(b)

| Number of logic levels | Number of circuits | Percent of circuits |
|---|---|---|
| 1 | 7 | 0.2 |
| 2 | 124 | 4 |
| 3 | 494 | 14 |
| 4 | 1798 | 51 |
| 5 | 792 | 23 |
| 6 | 288 | 8 |

(c)

| Class | Class description | Percent of circuits |
|---|---|---|
| A | Irreducible, not d-eulerian | 28 |
| B | Irreducible, d-eulerian | 11 |
| C | Reducible, d-eulerian | 45 |
| D | Reducible, not d-eulerian | 16 |

(d)

**Fig. 3.12.** Circuit statistics: (a) height distribution of circuits with $h \leq 4$; (b) minimum trail count distribution; (c) logic levels distribution; (d) classes based on reducibility and d-eulerian properties.

| R-TrailTrace | TrailTrace | | Uehara & vanCleemput [Uv81] | |
|---|---|---|---|---|
| Reduction in trail count | Number of circuits | Percent of circuits | Number of circuits | Percent of circuits |
| 0 | 1372 | 39 | 1714 | 49 |
| 1 | 2034 | 58 | 1436 | 41 |
| 2 | 94 | 3 | 340 | 10 |
| 3 | 2 | 0.06 | 12 | 0.3 |

(a)

| R-TrailTrace | TrailTrace | | Uehara & vanCleemput [Uv81] | |
|---|---|---|---|---|
| Percent reduction in area | Number of circuits | Percent of circuits | Number of circuits | Percent of circuits |
| 0 | 1372 | 39 | 1714 | 49 |
| 5-10 | 1824 | 52 | 346 | 38 |
| 11-19 | 298 | 9 | 420 | 12 |
| 20 or more | 8 | 0.2 | 22 | 1 |

(b)

| Circuit class | Percent of circuits | TrailTrace | R-Trail-Trace | Nair et al.[NB85] | Uehara & van Cleemput[Uv81] | Chen & Hou[CH88] |
|---|---|---|---|---|---|---|
| A | 28 | Optimal | Optimal | n/a | Nonoptimal | Nonoptimal |
| B | 11 | Optimal | Optimal | Optimal | Nonoptimal | Nonoptimal |
| C | 45 | Nonoptimal | Optimal | Optimal | Nonoptimal | Nonoptimal |
| D | 16 | Nonoptimal | Optimal | n/a | Nonoptimal | Nonoptimal |

| | |
|---|---|
| Optimal: | Guaranteed optimal |
| Nonoptimal: | Not guaranteed optimal |
| n/a: | Not applicable |

(c)

**Fig. 3.13.** Comparison of layout methods: (a) trail reduction; (b) area improvement; (c) optimality of methods for circuit classes defined in Fig. 3.12(d).

no layout at all for the 44% of circuits in classes A and D, which do not have a d-euler trail under any reordering.

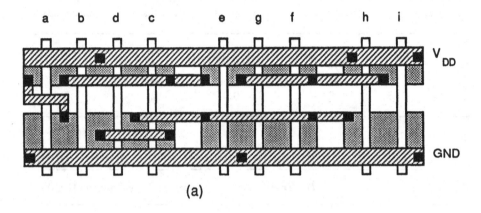

(a)

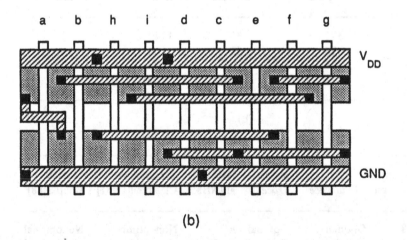

(b)

**Fig. 3.14.** Optimal layouts for the function (+/* (*/+ (+/*(*/+ih)(*/+gf)e)(+/* de)b)a) (a) before, and (b) after reordering.

The circuit corresponding to the layout of Fig. 3.14(a) is an example from the circuits considered (class C) that shows about a 20% area improvement due to reordering, compared to the optimal layout without reordering. The number of dual trails is reduced from three in Fig. 3.14(a) to one in Fig. 3.14(b). The composition expression for this example is

$$(+/* \; (*/+ \; (+/* \; (*/+ \; ih)(*/+ \; gf)e)(+/* \; de)b)a)$$

The resulting d-euler trail is *abhidcefg*. An example circuit from class D that *R-TrailTrace* reduces from four to two trails is the following:

$$(+/* \; (*/+ \; a(+/* \; bc)(+/* \; (*/+ \; de)(*/+fg)))(*/+ \; hi(+/* \; jk)(+/* \; lm)))$$

The area is reduced by 13% over that of *TrailTrace*. The reordering algorithm of Nair et al. is not applicable to circuits of this type. *R-TrailTrace* finds a single trail in the following circuit, for which the heuristic of [Uv81] finds four trails

$$(*/+ \; (+/* \; db)(+/* \; (*/+ \; d(+/* \; (*/+ \; de)(*/+ \; fg))(*/+ \; (+/* \; hi)(+/* \; (*/+ \; jk)(*/+ \; lm))))$$

The area improvement in this case is 21%.

We analyzed the execution time of *R-TrailTrace* for all 3503 circuits of height $h \le 4$. Figure. 3.15 shows the execution-time distribution for R-TrailTrace over all practical-sized circuits. The longest time is 87 s on a MicroVAX II, with the average time being about 25 s. Therefore, it is clear that even though *R-TrailTrace*'s execution time grows exponentially in the worst case, it is feasible for circuits of practical size. This is because the parameter in which the algorithm is exponential is not the number of transistors in the circuit, but the largest number of subgraphs represented by the children of any node in $T$. For example, the root node of $T$ in Fig. 3.3(a) has three children. The number of subgraphs in this arrangement never exceeds the circuit height $h$. Furthermore, isomorphic children of $T$ are not permuted. Therefore, the middle and right children of the root node of the tree in Fig. 3.3(a) are not permuted. The number of nonisomorphic subgraphs strictly in series or in parallel is even smaller, since a single edge (transistor) often constitutes a subgraph, and all single-edge subgraphs are isomorphic. Since isomorphic subgraphs are not permuted, execution time is further reduced. In the case of the class of circuits analyzed above, rarely does the total number of permutations exceed ten for the whole circuit; on average, it is about four. It is worth noting that, if desired, *R-TrailTrace* can be applied very efficiently to unusually large cells with $h \gg 4$, since many of them contain only a small number of different subgraphs strictly in series or in parallel.

| Execution time (s) | Number of circuits | Percent of circuits | Cumulative percent |
|---|---|---|---|
| 0-1 | 13 | 0.3 | 0.3 |
| 1-10 | 391 | 11 | 11.3 |
| 10-20 | 1080 | 31 | 42.3 |
| 20-30 | 910 | 26 | 68.3 |
| 30-40 | 573 | 16 | 84.3 |
| 40-50 | 301 | 9 | 93.3 |
| 50-60 | 145 | 4 | 97.3 |
| 60-70 | 64 | 2 | 99.3 |
| 70-80 | 16 | 0.5 | 99.8 |
| 80-90 | 10 | 0.2 | 100.0 |

**Fig. 3.15.** Execution time distribution for *R-TrailTrace* on all practical-sized circuits.

Finally, we note that using our layout theory [MH87], Huang and Sarrafzadeh have shown that the major steps of our layout algorithms can be efficiently parallelized [HS88].

## 3.6 MINIMUM-WIDTH ROWS OF CELLS

We discuss next how *TrailTrace* can be augmented to solve the WM (width minimization for multiple cells) problem exactly and similarly how *R-TrailTrace* can be modified to solve exactly the WRM (width minimization with reordering for multiple cells) problem. When used as components of a larger design, the functional cells generated by either of our algorithms are placed side-by-side with common horizontal $V_{DD}$ and GND buses. In order to avoid a diffusion gap between neighboring cells, the trail-cover type for each functional cell can be chosen such that its $V_{DD}$ and GND contact terminals are exposed on one side of the cell. This allows adjacent cells to be connected by diffusion abutment, thus saving area. The $V_{DD}$ and GND terminals can be placed on one side of a cell by a careful choice of the trail cover from the final CTC of the entire graph. This choice is made in the final step of the main procedure of the algorithm (statement 1 in Fig. 3.9). Since the S terminal of $M$ corresponds to GND and the N terminal in $M^d$ corresponds to $V_{DD}$, we can choose a trail cover that contains endpoints with labels (S,N), such as (S,N)/(S,S). If several trail covers with the same number of trails exist, we can choose the one with (S,N). There is little advantage in selecting a trail cover from the CTC that contains (S,N) but has more than the minimum number of trails, because the most that can be gained by one cell abutment is half a diffusion gap per cell. Ideally, if we had total freedom to place the $n$ cells anywhere in the row, pairs of such cells would be placed side-by-side, thus reducing the row width by $n/2$ gaps if $n$ is even, and by $(n-1)/2$ if $n$ is odd, and resulting in the absolute minimum width. In practice, when cells are preplaced horizontally based on other design criteria such as routability, neighboring cells can be paired and abutted similarly to yield the minimum-width row achievable for that cell placement. In Chapter 6, we present an algorithm *HRM-TrailTrace* that constructs cell arrays that are of minimum height as well as width.

# CHAPTER IV

# PLANAR CELL WIDTH MINIMIZATION

In Chapter 3, we presented an exact width minimization method for the layout of dual series-parallel circuits, which are a subset of the class of dual planar circuits. In this chapter, we generalize that method to apply to a special class of nonseries-parallel circuits, thus relaxing assumption 2 of the Uehara-vanCleemput layout style described in Fig. 2.1. The circuits in question are those that can be modeled by planar graphs, that is, graphs that can be drawn in a plane with no edges crossing. Planar graphs constitute the largest class of graphs that have graphical duals [Ha69]. We develop a new algorithm *P-TrailTrace* to handle any static CMOS circuit that can be modeled as a dual pair of two-terminal planar multigraphs (TTPMs), including both the series-parallel and nonseries-parallel cases.

Relatively little is known about nonseries-parallel graphs, in marked contrast to the series-parallel class. Nevertheless, circuit designers use nonseries-parallel circuits to reduce layout area and improve performance. Figures 4.1 and 4.2 illustrate the area savings that nonseries-parallel circuits may offer compared to series-parallel circuits. Figures 4.1(a) and (b) show graph models of nonseries-parallel and series-parallel implementations of the function

$$z = \neg(*(+ab)(+cd)(+ef)(+g(*(+ad)(+bc)(+h(*(+be)(+af))) (+h(*(+ae)(+cf))))$$

respectively; Figs. 4.2 (a) and (b) show the corresponding layouts in our functional cell style. The series-parallel circuit layout is 2.3 times larger than the nonseries-parallel one. Our goal is a practical layout method that solves the W layout problem exactly for nonseries-parallel dual CMOS circuits.

This chapter is organized as follows. A generalized graph composition method is presented in Section 4.2, and our extension of the theory of dual trail covering is discussed in Section 4.3. We present the *P-TrailTrace* algorithm and discuss its properties in Section 4.4. Section 4.5 reports the results of a complete study of all practical-sized nonseries-parallel planar circuits using *P-TrailTrace*.

## 4.1  NONSERIES-PARALLEL COMPOSITION

We present two main extensions to the series-parallel cell theory of Chapter 3. First, we extend the graph composition method to handle planar graphs; note that the composition process affects the efficiency, not the optimality, of the layout algorithm. We also generalize the trail covering theory to generate minimum covers for planar graphs; this is the basis for the width-minimization portion of the algorithm.

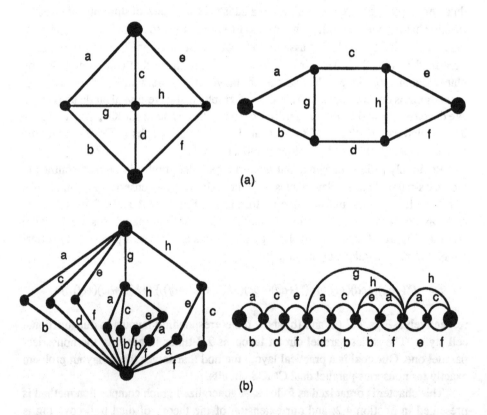

**Fig. 4.1.** (a) Nonseries-parallel and (b) series-parallel graph models for circuits implementing the function $z = \neg(*(+ab)(+cd)(+ef)(+g(*(+ad)(+bc)(+h(*(+be)(+af)))(+h(*(+ae)(+cf))))$.

### 4.1.1 Graph Composition

As with the series-parallel layout algorithms of Chapter 3, the layout method presented here is based on composing graphs from their single-edge subgraphs. This composition process guides graph-cover concatenation, as in the previous chapter. The complete set of dual trail covers is known for the trivial case and is the same as the set of covers used for series-parallel graphs. In the latter case, composition directly follows in bottom-up fashion a composition tree $T$ derived naturally from the structure of the logical function that the circuit implements. A more general method is needed to derive $T$ for nonseries-parallel dual planar graphs, since $T$ for such graphs cannot easily be derived from the structure of their logical expression. This

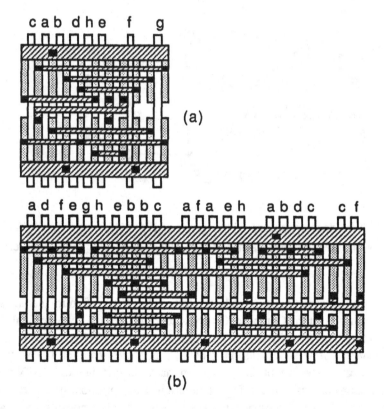

**Fig. 4.2.** Layouts corresponding to: (a) the nonseries-parallel and (b) the series-parallel graphs of Fig. 4.1.

generalization also requires new composition operators in addition to the two operators */+ and +/* that characterize the composition of series-parallel graphs.

First, we develop a composition algorithm Compose_Graph for nonseries-parallel graphs and then discuss the composition operators.

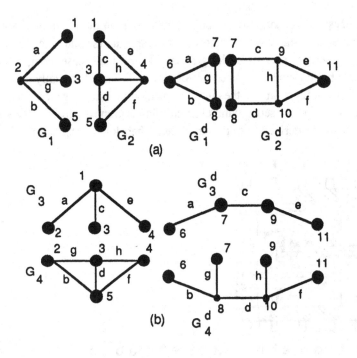

(a)

(b)

**Fig. 4.3.** Two decompositions of the graph of Fig. 4.1(a) into: (a) $G_1/G_1^d$ and $G_2/G_2^d$ with three terminals each, and (b) $G_3/G_3^d$ and $G_4/G_4^d$ with four terminals each.

**Graph Composition Algorithm.** The goal of our planar graph composition algorithm Compose_Graph is to compose any two-terminal planar graph and its dual from its edges. However, although TTSPMs can be composed solely from TTSPM subgraphs, in general, a nonseries-parallel TTPM requires component graphs with more than two terminals. In fact, unlike TTSPMs which have a unique composition tree $T$ for two-terminal composition, nonseries-parallel TTPMs have alternative composition trees resulting in components with various numbers of designated terminals. For instance, consider the two-terminal dual graphs in Fig. 4.1(a); terminals in the figures are drawn larger in size than nonterminals. Each graph can be partitioned into the subgraphs $G_1/G_1^d$ and $G_2/G_2^d$ of Fig. 4.3(a) with three terminals, or into $G_3/G_3^d$ and $G_4/G_4^d$ of Fig. 4.3(b) with four terminals.

Therefore, we incorporate a heuristic in Compose_Graph that attempts to find a composition of a TTPM that minimizes the maximum number of terminals of any subgraph in the set of component subgraphs of $G$ over all steps in the composition process. This heuristic is incorporated in the algorithm because the number of trail covers types is directly related to the number of terminals in the resulting subgraph. The number of covers types affects the worst-case run time of the trail-concatenation algorithm *P-TrailTrace*, but does not affect the width of the layouts. Therefore, the use of this heuristic in no way affects the correctness of the algorithm, but only attempts to improve the efficiency of *P-TrailTrace*.

Compose_Graph has polynomial time complexity. We show in Section 4.3 that the efficiency heuristic in the composition algorithm actually minimizes the maximum number of terminals over the entire composition process for all the practical-sized planar circuits.

Compose_Graph is given in Fig. 4.4. In the first **for** loop the procedure Find_best_graph_to_merge_with pairs each edge $G_i$ of $G$ with some other edge $G_j$ according to the following two selection rules corresponding to statements 1 and 2 in Fig. 4.4.

Selection rule 1:    Pair graph $G_i$ having $T_i$ terminals with a graph $G_j$ having $T_j$ terminals, resulting in the composed graph $G_k$ with $T_k$ terminals such that $T_k$ is minimized.

Selection rule 2:    Pair graph $G_i$ having $T_i$ terminals with a graph $G_j$ having $T_j$ terminals, resulting in the composed graph $G_k$ with $T_k$ terminals such that the quantity $\Delta T = T_k - (T_i + T_j)$ is minimized.

The first rule is to compose all graphs resulting in $n$-terminal graphs before composing any graphs resulting in $n + 1$ terminals, starting with $n = 2$. This implies that the series-parallel components of a TTPM are composed before the nonseries-parallel parts, because the series-parallel parts always have two terminals. Therefore, Compose_Graph can be used to identify the series-parallel components of a TTPM. This capability may be used to apply the reordering algorithm for series-parallel TTSPMs, if so desired, by permuting the order of such two-terminal components. The second rule composes graphs that result in the largest decrease in the number of terminals. This can produce the most dramatic reductions in the size of the resulting CTC compared to the sizes of the CTCs of $G_1$ and $G_2$. The selection rules and their relative priorities are chosen to serve one overall goal: to

```
         procedure Find_best_graph_to_merge_with(G);
         begin
         T_min := ∞;   ΔT_min := ∞;
         for each component graph C sharing terminals with G do
               begin
1:             if (T_min > T_k ) where graph G with T_G terminals composes a graph C with T_C
               terminals resulting in the graph G_k with  T_k terminals
2:                 or ( (T_min = T_k ) and  (ΔT_min > T_k  - (T_G + T_C ) ) ) then
                   begin
                   Pair graphs G and C;
                   TG_1 := T_k ;  ΔTG_1 := T_k  - (T_G + T_C);
                   end;
               end;
         end; {Find_best_graph_to_merge_with}

         procedure Find_best_pair_of_graphs_to_merge(G_1,G_2);
         begin
         T_min := ∞;   ΔT_min := ∞;
         for each component graph G_1 do
               begin
               if (T_min > TG_1 ) or ((T_min = TG_1) and  (ΔT_min > ΔTG_1 )) then
                   begin
                   G_2 := the pair of  G_1;
                   T_min := TG_1 ;   ΔT_min := ΔTG_1 ;
                   end;
               end;
         end; {Find_best_pair_of_graphs_to_merge}

         procedure Compose_Graph;
         begin
         Input_graph(G);
         for each Edge of graph G do
               Find_best_graph_to_merge_with(Edge);
         while (Graph_count > 1) do
               begin
               Find_best_pair_of_graphs_to_merge(G_1,G_2);
               Compose_graph_G_3(Terminals_3, G_1,G_2);
               Find_best_graph_to_merge_with(G_3);
               for each graph G whose best neighbor is G_1 or G_2 do
                    Find_best_graph_to_merge_with(G);
               Decrement Graph_count;
               end; {while}
         end; {Compose_Graph}
```

**Fig. 4.4.** The graph composition algorithm Compose_Graph.

minimize the maximum number of terminals of any component graph in the graph composition process. These rules were developed and refined through experimentation with the composition of practical-sized graphs. The quality of the rules can be evaluated by the fact that they allow no component of any practical-sized graph to exceed three terminals in the composition process. This important result is discussed more fully in Section 4.3.2.

In the first step of the **while** loop of Compose_Graph, the procedure Find_best_pair_of_graphs_to_merge selects the pair of graph components $G_1$, $G_2$ to compose next based on which pair best meets the two selection rules. The selected pair is composed by the procedure Compose_graph_$G_3$ according to the terminals $G_1$ and $G_2$ share. The composed graph $G_3$ is paired with another component by Find_best_graph_to_merge_with as described above. The graphs that had been paired with either $G_1$ or $G_2$ are similarly paired with other components, since $G_1$ and $G_2$ no longer exist as independent components after being composed. Finally, the graph count is decremented and the **while** loop is repeated until the entire $G$ is composed.

We illustrate Compose_Graph using the graphs in Fig. 4.5. The graph $G_9$ and its dual $G_9{}^d$ of Fig. 4.5(f) are decomposed into the two six edge graphs in Fig. 4.5(a). First, the edges are paired according to the two selection rules. The first rule pairs graphs that result in two-terminal graphs, if possible. This rule pairs edge graphs $e$ and $f$ of Fig. 4.5(a), because they are the only pair which, when composed, results in a two-terminal graph. All other pairings result in three-terminal graphs; therefore, we also pair edge $a$ with $b$, and $c$ with $d$. The **while** loop of Fig. 4.4 is entered with a graph count of six, i.e., each of the six edges of the graph is a component. First, the pair of graphs that best meets the selection rules is chosen for composition from the pairs produced above. Pair $e$ and $f$ is selected and composed to form the two-terminal graphs $G_1/G_1{}^d$ in Fig. 4.5(b), with terminals 3 and 4 for $G_1$ and 7 and 9 for its dual. Next $G_1/G_1{}^d$ is paired with edge $d$, although it could be paired with either edge $b$ or $c$ because the composed graph has three terminals in any case, whereas pairing it with edge $a$ results in a four-terminal graph. Since no other edges were paired with edges $e$ or $f$, we decrement Graph_count to five and repeat the **while** loop. On the second iteration of the loop, the graphs $G_1/G_1{}^d$ and $G_2/G_2{}^d$ of Fig. 4.4(b) are selected to compose for the following reasons. No two graphs can combine to form a two-terminal graph, and several pairs result in three-terminal graphs; therefore, selection rule 1 results in a tie. Rule 2 selects graph $G_1/G_1{}^d$ to compose with graph $G_2/G_2{}^d$ resulting in $G_3/G_3{}^d$ of Fig. 4.5(c), because $\Delta T = T_{G_3} - (T_{G_1} + T_{G_2}) = 3 - (3 + 2) = -2$. Any other graphs $G_i, G_j$ yield $\Delta T = T_{ij} - (T_i + T_j) = 3 - (2 + 2) = -1$. The process continues, composing the graph in successive

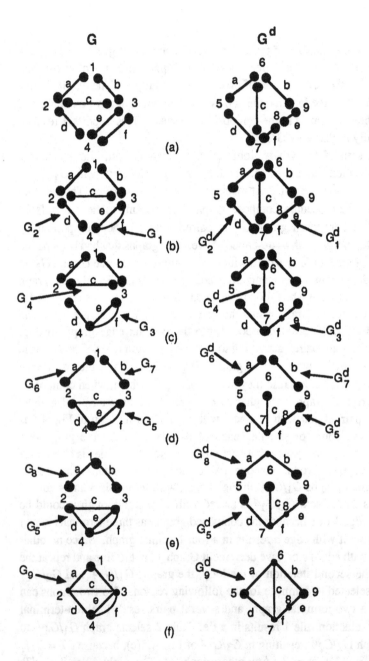

**Fig. 4.5.** Various stages in the composition of dual planar graphs by the Compose_Graph algorithm.

iterations. In Fig. 4.5(c), $G_3/G_3{}^d$ composes with $G_4/G_4{}^d$ to form $G_5/G_5{}^d$ in Fig. 4.5(d). $G_6/G_6{}^d$ composes with $G_7/G_7{}^d$ to form $G_8/G_8{}^d$ in Fig. 4.5(e), and finally, $G_5/G_5{}^d$ composes with $G_8/G_8{}^d$ to form $G_9/G_9{}^d$ in Fig. 4.5(f), completing the composition process. Note that no component $G_i/G_i{}^d$ generated in this example has more than three terminals.

The time complexity of Compose_Graph is $O(E^3)$ to compose an entire graph, where $E$ is the number of edges in the graph $G = (V,E)$. Each composition step takes $O(E^2)$ time, and requires one iteration of the while loop of Fig. 4.4. This while loop is executed $E-1$ times.

**Planar Graph Composition Operators.** Graph operators like the series-parallel operations +/* and +/* combine terminals of graphs in two different ways we term connecting and merging. Two terminals from different graphs are *connected* if, on composing the graphs and joining the terminals into a single vertex according to some composition operator, the resulting vertex is a designated terminal. If the resulting vertex is not a terminal, then the terminals are said to be *merged*. A connection operation $C_i\{v_1,...,v_i\}$ is used in a graph composition step to connect $i$ terminals $\{v_1,...,v_i\}$ in $G_1$ with $i$ terminals $\{v_1,...,v_i\}$ in $G_2$, if either all edges incident with the vertices $\{v_1,...,v_i\}$ in the final graph are not incident with them in the graph $G_3$ resulting from the composition, or if the vertices are terminals in the final graph. A merge operation $M_i\{v_1,...,v_i\}$ is used to merge $i$ terminals $\{v_1,...,v_i\}$ in $G_1$ with $i$ terminals $\{v_1,...,v_i\}$ in $G_2$ if, as a result of the operation, all edges incident with the vertices $\{v_1,...,v_i\}$ in the final graph are incident with them, and these vertices are not terminals in the final graph.

The */+ operation composes two TTSPMs in series (*) with a single terminal of each graph being merged using the $M_1$ operation. Both pairs of terminals are connected in the dual pair of graphs, which are composed in parallel (+) by the $C_2$ operation. Thus, in the notation introduced above, the */+ operation is denoted $(M_1/C_2)$. The +/* operator applies the same connection and merge operations, except that it reverses the roles of the graphs and their duals, and so is denoted $(C_2/M_1)$.

In general, composition operators apply a combination of connection and merge operations to a pair of graphs with an unrestricted number of terminals. Therefore, we generalize our previous set of graph composition operators to include general connection and merge operations. We denote a *composition operator K* by $(M_n,C_m/M_p,C_q)$, where $M_i$ indicates a merge operation and $C_i$ denotes a connection operation; $n, m, p$ and $q$ denote the number of terminal pairs involved in the respective operations. The first two operations, $M_n$ and $C_m$ are applied to the two graphs $G_1$ and $G_2$ to be composed, and the second two, $M_p$ and $C_q$ to their duals $G_1{}^d$ and $G_2{}^d$. If a value of $n, m, p,$ or $q$ in an operator is zero, then the corresponding connection or merge operator is dropped from the notation. For example, the full notation for the class 1 operator actually is $(M_1,C_0/M_0,C_2)$, but we drop the $C_0$, and $M_0$ terms, resulting in the simpler form $(M_1/C_2)$. We will sometimes identify the vertices to which $M_i$ or $C_j$ apply by listing them in parentheses following the operator. We always label terminals to be connected or merged by the same label. For example, the expression $G_1$ $(M_1\{3\}/C_2\{1,4\})$ $G_2$ denotes the merging of one pair of terminals labeled 3 in both $G_1$ and $G_2$, and connecting of two pairs of terminals, 1 and 4 in $G_1^d$ and $G_2^d$.

| Operator class | No. of terminals in $G_1$ | No. of terminals in $G_2$ | No. of terminals in $G_3$ | Operator $K$ | Dual operator $K^d$ |
|---|---|---|---|---|---|
| 1 | 2 | 2 | 2 | $(M_1/C_2)$ | $(C_2/M_1)$ |
| 2 | 2 | 2 | 3 | $(C_1/C_1)$ | $(C_1/C_1)$ |
| 3 | 2 | 3 | 3 | $(M_1/C_2)$ | $(C_2/M_1)$ |
| 4 | 3 | 3 | 3 | $(M_1,C_1/M_1,C_1)$ | $(M_1,C_1/M_1,C_1)$ |
| 5 | 2 | 3 | 2 | $(M_1,C_1/M_1,C_1)$ | $(M_1,C_1/M_1,C_1)$ |
| 6 | 3 | 3 | 2 | $(M_2/M_1,C_2)$ | $(M_1,C_2/M_2)$ |

**Fig. 4.6.** The composition operators for all three-terminal planar dual graphs.

Figure 4.6 lists all possible composition operations for the set of two- and three-terminal dual planar graphs. Figure 4.7 illustrates the following typical composition operation in class 3 of Fig. 4.6. $G_3/G_3^d$ is composed from $G_1/G_1^d$ and $G_2/G_2^d$ via an $(M_1/C_2)$ operation in which one pair of terminal of the graphs, labeled 2, is merged and two pairs of terminals, labeled 5 and 6, in the dual graphs are connected. The composition expression for this example is written as follows:

$$G_3/G_3^d = G_1/G_1^d \ (M_1\{2\}/C_2\{5,6\}) \ G_2/G_2^d.$$

Figure 4.7 shows that terminal 2 in $G_1$ and terminal 2 in $G_2$ are merged, becoming a nonterminal in $G_3$, and terminals 5 and 6 in $G_1^d$ and $G_2^d$ are connected, thus they remain terminals in $G_3^d$. This operation, therefore, composes a two- and a three-terminal graph into a three-terminal graph. Another example of a composition operation is the one in Fig. 4.8, which composes two three-terminal graphs resulting in a two-terminal graph, as indicated in Fig. 4.6. $G_3/G_3^d$ is composed from $G_1/G_1^d$ and $G_2/G_2^d$ via an $(M_2/M_1,C_2)$ operation in which two pairs of terminal of the graphs labeled 3 and 4 are merged and one pair of terminals labeled 7 in the dual graphs is merged, with two pairs of terminals, labeled 6 and 8, being connected.

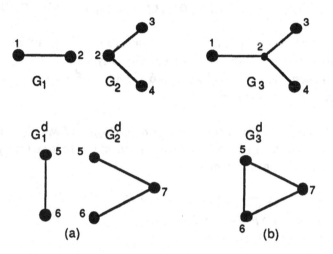

**Fig. 4.7.** The composition $G_3 / G_3^d = G_1 / G_1^d$ (M₁{2} / C₂{5,6}) $G_2 / G_2^d$ using a class 3 operator: (a) before composition; (b) after composition.

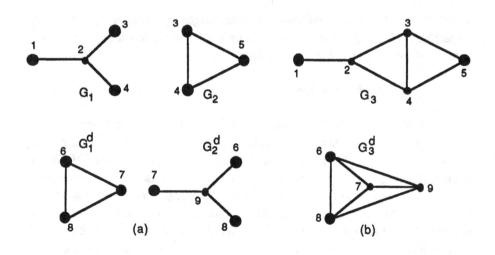

**Fig. 4.8.** The composition $G_3 / G_3^d = G_1 / G_1^d$ (M₂{3,4} / M₁{7},C₂{6,8}) $G_2 / G_2^d$ using a class 6 operator: (a) before composition; (b) after composition.

The composition expression in this case is written as

$$G_3/G_3{}^d = G_1/G_1{}^d \ (M_2\{3,4\} \ / \ M_1\{7\}, \ C_2\{6,8\}) \ \ G_2/G_2{}^d.$$

Terminals 3 and 4 in $G_1$ and in $G_2$ are merged in Fig. 4.8(a), becoming nonterminals in $G_3$, as shown in Fig. 4.8(b). Similarly, terminal 7 in $G_1{}^d$ and $G_2{}^d$ becomes a nonterminal in $G_3{}^d$. Terminals 6 and 8 in $G_1{}^d$ and $G_2{}^d$ are connected, thus remaining terminals in $G_3{}^d$.

**Lemma 4.1:** Every planar composition operator $(M_n, C_m \ / \ M_p, C_q)$ must satisfy the following three constraints, where the two operand graphs have $i$ and $j$ terminals and the composed graph has $k$ terminals:

1. $k = i + j - (2n + m) = i + j - (2p + q).$
2. $n + m - 1 = p$ and $p + q - 1 = n.$
3. $\max\{n+m,p+q\} \leq \min\{i,j\}.$

**Proof:** The basis for these constraints is Euler's formula

$$V - E + F = 2$$

where $V, E$ and $F$ are the number of vertices, edges and faces of a planar graph, respectively [Ha69]. In the case of graphs with $T$ terminals, where the external face is partitioned into $T$ faces, the formula becomes

$$V - E + F = T + 1.$$

The first constraint is that a planar graph $G$ and its dual $G^d$ must have the same number of terminals. This follows from Euler's formula applied to both graphs, where the $V$ vertices in $G$ correspond to the $F$ faces in $G^d$, and vice versa. If $G$ has $T$ terminals, then $G^d$ has $T$ external faces, and vice versa. In order for $G_3$ and $G_3{}^d$ to have the same number of terminals, the number of external faces in $G_3$ created by the operation must equal the number of new terminals created in $G_3{}^d$, and vice versa. Each merge operation reduces the total number of terminals by two, and connection reduces it by one. If $G_1$ has $a$ terminals and $G_2$ has $b$ terminals, and $G_3$ has $d$ terminals, then $d = a + b - (2n + m) = a + b - (2p + q)$, as required. Note that this also implies that $2n + m = 2p + q$.

The second constraint is that number of internal faces in $G_3$ created by the composition operator must equal the number of nonterminals created by merges in $G_3{}^d$, and vice versa. In terms of Euler's formula, if $G$ has $T$ terminals, then $G^d$ has

$V - T$ internal faces, and vice versa. This means that $n + m - 1 = p$ and $p + q - 1 = n$. Each nonterminal in G corresponds to a unique internal face in $G^d$, and vice versa. There are $n + m - 1$ internal faces created in $G_3$ by $n$ merge and $m$ connection operations, and $p + q - 1$ internal faces created in $G_3^d$ by $p$ merge and $q$ connection operations. The $n$ merge operations create $n$ nonterminals in $G_3$ and the $p$ merge operations in the dual graphs create $p$ nonterminals in $G_3^d$. The number of internal faces in $G_3$ must equal the number of nonterminals in $G_3^d$, and vice versa. Therefore, $n + m - 1 = p$ and $p + q - 1 = n$.

The last constraint states that the number of operations in the operator must be less than or equal to the number of terminals of the graph with the fewer terminals, i.e., $\max\{n+m,p+q\} \leq \min\{a,b\}$. Each merge and connection operation involves one terminal from $G_1$ ($G_1^d$) and one from $G_2$ ($G_2^d$). The maximum number of such operations cannot exceed the number of terminals in the graph with the smaller number of terminals; otherwise, one of its terminals is involved in more than one operation, which is impossible. $\square$

The three constraints of Lemma 4.1 lead to the following result.

**Theorem 4.1.** There are exactly nine nonequivalent composition operator of the form $(M_n,C_m/M_p,C_q)$ for the set of two- and three-terminal dual planar graphs.

**Proof:** The proof is by exhaustive enumeration. This method of proof is feasible here because the values for $i, j$ and $k$ in the three constraints of Lemma 4.1 are limited to two and three. We generate all possible values for $m, n, p$ and $q$ that satisfy the three constraints with the above values for $i, j$ and $k$. These values correspond to the nine composition operators in Fig. 4.6. The same values for $m, n, p$ and $q$ corresponding to different value of $i, j$ and $k$ result in nonequivalent composition operators. $\square$

Figure. 4.6 groups the nine planar composition operators into six classes according to the number of terminals of the operand and result graphs. Examples of the same values for $m, n, p$ and $q$ corresponding to different value of $i, j$ and $k$ are found in the operator $(M_1/C_2)$ in class 1 and $(M_1/C_2)$ in class 3 of Fig. 4.6. As noted above, the dual operators in class 1 of Fig. 4.6 are the */+ and +/* operators for series-parallel graphs. The remaining five classes contain new operators used to compose two- and three-terminal planar dual graphs.

Reversing the role of the operations on $G$ with those on $G^d$ in a composition operator $K$ results in a *dual operator* $K^d$. It is interesting that for some of the classes in Fig. 4.6, such as class 1, $K \neq K^d$ whereas for other classes, such as class 2, $K = K^d$. Figures 4.7 and 4.8 above illustrate the nonself-dual composition operators. Figures 4.9 - 11 illustrate self-dual operators. In Fig. 4.9, the $(M_1/C_2)$ operator composes 2 two-terminal graphs into a three-terminal graph. This operator is the first operator not in class 1 to be used by Compose_Graph after all TTSPM components are composed. Figure 4.10 shows the operator $(M_1,C_1/M_1,C_1)$ composing three-terminal graphs into a three-terminal graph. Except for the series-parallel operators of class 1, this is the only operator whose domain and codomain are the same, i.e., the set of three-terminal planar graphs. Figure 4.11 illustrates the operator $(M_1,C_1/M_1,C_1)$, which has the same values of $m$, $n$, $p$ and $q$ as the operator of Fig. 4.10, but whose domain is the set of two- and three-terminal graphs, and whose codomain is the set of two-terminal graphs.

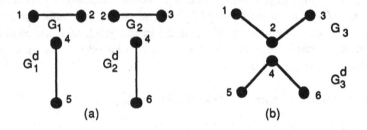

(a)                                              (b)

**Fig. 4.9.** The composition $G_3/G_3{}^d = G_1/G_1{}^d$ $(C_1(2)/C_1(4))$ $G_2/G_2{}^d$ using a class 2 operator: (a) before composition; (b) after composition.

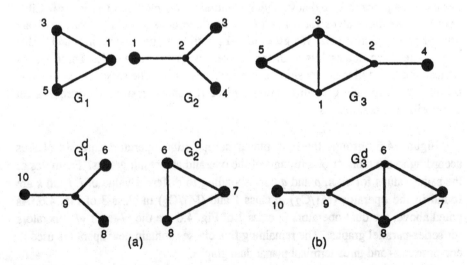

**Fig. 4.10.** The composition $G_3/G_3{}^d = G_1/G_1{}^d$ $(M_1(1),C_1(3)/M_1(6),C_1(8))$ $G_2/G_2{}^d$ using a class 4 operator: (a) before composition; (b) after composition.

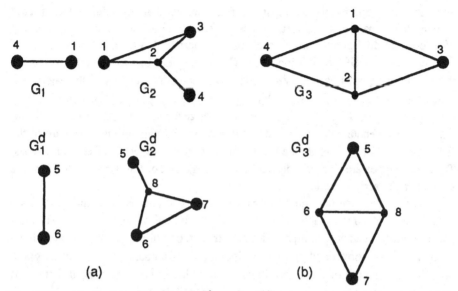

**Fig. 4.11.** The composition $G_3/G_3{}^d = G_1/G_1{}^d$ $(M_1\{1\},C_1\{4\} / M_1\{6\},C_1\{5\})$ $G_2/G_2{}^d$ using a class 5 operator: (a) before composition; (b) after composition.

Our dual trail-covering algorithm *P-TrailTrace* for planar graphs receives the circuit in the form of a list of dual graph edges with no explicit composition operations. The operators are applied to the components of a graph by Compose_Graph. The particular operators and the order in which they are applied to the components of the circuit are determined by the structure of the circuit under consideration and the goal of minimizing the number of terminals of the components.

## 4.1.2 Trail Covering

In Chapter 3, we developed a theory of dual trail covering for two-terminal series-parallel multigraphs, which provided the basis for the two layout algorithms *TrailTrace* and *R-TrailTrace*. Here we generalize that theory to include nonseries-parallel planar graphs as well. Parts of the earlier theory directly apply here without modification. Trail covers of TTPMs consist of distinguished and/or internal dual trails. The trails in the covers of two graphs can be concatenated if they are compatible, where compatibility is defined as before, but over the new set of graph planar composition operators.

Other aspects of the theory require generalization to accommodate nonseries-parallel graphs. The vertex types of a TTSPM are N, S and I. The vertex types of an $n$-terminal graph are $V_1, ..., V_n$ and I. Endpoint types are now defined as ordered pairs on this expanded set of vertex types. There are $n^2 + 1$ endpoint types for a pair of $n$-terminal dual graphs, where each of the $n$ terminal types of $G$ may be paired with each one of $G^d$, in addition to the (I,I) type. Let the set of vertex types for three-terminal graphs be $\{V_1, V_2, V_3, I\}$. The ten endpoint types of three-terminal graphs are $\{(V_1,V_1), (V_1,V_2), (V_1,V_3), (V_2,V_1), (V_2,V_2), (V_2,V_3), (V_3,V_1), (V_3,V_2), (V_3,V_3), (I,I)\}$. The first nine characterize a pair of endpoints of a dual trail that can concatenate with other trails. All nonterminal pairs of endpoints are of the (I,I) type. Trail types are defined as before but on the larger set of endpoint types. Trail-cover types are similarly defined on this larger set of trail types. A CTC is the same as in the series-parallel case.

We now describe how dual trails in the covers of two pairs of dual graphs are concatenated when the graphs are composed. Assume the case where the terminals corresponding to an endpoint pair of a dual trail in one of the graph pairs are joined to the terminals corresponding to an endpoint pair of a dual trail in the other graph pair. If one or both of those terminals of an endpoint pair are merged instead of connected, then the two trails are concatenated in the resulting trail cover. Otherwise, they remain separate trails in the new cover. Whether terminals are merged or connected is determined by the structure of the graphs and the graph composition operation. Some of the trails in the graphs become isolated from the terminals of the new graph as a result of the composition of the graphs. These internal trails are retained in the trail cover, but cannot participate in any further concatenation.

We applied Compose_Graphs to all nonseries-parallel planar circuits with height $h \leq 4$ and found that only two- or three-terminal planar graphs resulted at each stage in their composition. We also examined several graphs with $h > 4$ and found that no more than four terminals were required.

In Chapter 3, we showed that two-terminal dual series-parallel graph are characterized by $N = 42$ trail cover types. The size $N$ of the complete set of cover types grows with the number of terminals of a graph, because the maximum number of covers of each graph of a pair to be composed is limited by $N$. The time of the composition operation is $N^2$ times longer than the time for the concatenation of a single pair of covers. Therefore, the overall time to concatenate the trail sets of all the subgraphs is $O(EN^2)$, where $E$ is the number of edges in $G$. We had previously calculated exactly the set of trail cover types for dual TTSPMs, which are reported in Chapter 3; see Fig. 3.8. We have calculated $N = 429$ to be the number of trail cover types for any pair of three-terminal dual planar multigraphs by using this same basic method. We wrote a program to perform the calculations, based on the algorithm Compose_Graph in Fig. 4.4. We start with a set of the two primitive cover types for a single-edge TTSPM and its dual, and alternately apply the */+ and +/* series-parallel operators in class 1 of Fig. 4.10 to each pair of cover types in this set, adding any new two-terminal types to the set. This is repeated until closure for the two-terminal types is achieved. Then the operators of class 2 are applied, generating three-terminal cover types. The operators in classes 3 and 4 are alternately applied until closure is achieved. Class 5 and 6 operators are used to attempt to generate any new two-terminal types. This whole cycle is repeated until closure over two- and three-terminal cover types is found. The resulting number of two-terminal cover types is 42 and three-terminal cover types is 429. The number of different three-terminal cover types provides an upper bound on the number of covers in a CTC at any stage in the graph covering process. This bound allows us to more accurately determine the worst-case run time of the *P-TrailTrace* algorithm described below.

## 4.2  P-TRAILTRACE ALGORITHM

*P-TrailTrace* solves the W layout problem for dual planar circuits. It finds all minimum-sized trail cover types of the graph, and selects one cover with the minimum number of dual trails. This cover is used to construct the minimum-width layout.

### 4.2.1  Description

We now give an overview of *P-TrailTrace*, which is summarized in Fig. 4.12. Compose_Graph (Fig. 4.4) is an integral part of *P-TrailTrace*. Initially, the circuit graph is entered as a list of pairs of dual graph edges. Each edge-graph is paired with another according to the two selection rules described in Section 4.1.1. The first rule pairs graphs whose composed graph has fewest terminals $T$, and the second rule breaks ties resulting from the first rule by pairing graphs whose composed graph minimizes $\Delta T$. The **while** loop of Fig. 4.12 is repeated until the graph is fully composed. In each iteration of this **while** loop, the following are accomplished. The pair of component graphs that best meets the two above-mentioned rules is selected. The graphs are composed and the list of terminals for the resulting graph is computed. This guides the CTC concatenation process, which immediately follows. The concatenation of the CTCs of the components is accomplished by the pairwise concatenation of the covers of the CTC lists. The resulting graph $G_3$ is paired with another component graph according to the selection criteria. The graphs previously paired with $G_1$ and $G_2$ are similarly paired, since $G_1$ and $G_2$ no longer exist independently. Finally, the graph count is decremented. When the **while** loop is exited, the smallest trail cover in the CTC of the entire graph is chosen as the output of the algorithm. This, in turn, is used to construct the minimum-width layout of the circuit.

```
procedure Concatenate_CTC( var Terminals, CTC, CTC_1, CTC_2);
begin
     for each trail cover TC_1 in CTC_1 do
          for each Trail Cover TC_2 in CTC_2 do
               begin
               Create_new_TC;
               Concatenate_TC( Terminals, newTC, TC_1, TC_2);
               Insert_TC_into_CTC( newTC, CTC);
               end;
end; {Concatenate_CTC}

procedure P-TrailTrace;
begin
Input_graph(G);
for each Edge of graph G do
     Find_best_graph_to_merge_with(Edge);
while (Graph_count > 1) do
     begin
     Find_best_pair_of_graphs_to_merge(G_1, G_2);
     Compose_graph_G_3(Terminals_3, G_1, G_2);
     Concatenate_CTC(Terminals_3, CTC_3,CTC_1, CTC_2);
     Find_best_graph_to_merge_with(G_3);
     for each graph G whose best neighbor is G_1 or G_2 do
          Find_best_graph_to_merge_with(G);
     Decrement Graph_count;
     end; {while}
Output_CTC;
end; {P-TrailTrace}
```

**Fig. 4.12.** An outline of *P-TrailTrace*.

The Concatenate_CTC procedure is now described more fully. The procedure pairs each trail cover $TC_1$ of $CTC_1$ with every $TC_2$ of $CTC_2$. It calls the procedure Concatenate_TC on each of these pairs $TC_1$ and $TC_2$ along with the terminals. Concatenate_TC duplicates the two trail covers and concatenates all compatible trails; the resulting set of trails is the new TC. This new TC is inserted into CTC, the list of trail covers resulting from the concatenation operations. The procedure

Insert_TC_into_CTC adds the new TC to the CTC only if there is no smaller TC of its cover type presently on the CTC list.

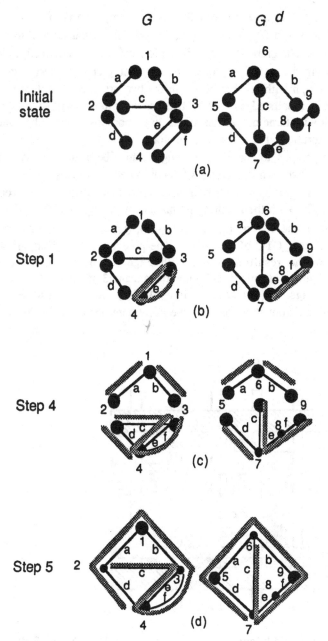

**Fig. 4.13.** Composition of a dual graph $G/G^d$.

## 4.2.2  Design Example

We illustrate the *P-TrailTrace* algorithm by an example based on a circuit with the graph model shown in Fig. 4.5; this graph previously was used to illustrate the composition process. The graph is entered in the form of six dual graph edges labeled *a* through *f* as in Fig. 4.13(a). The procedure Find_best_graph_to_merge_with pairs each component graph with another based on the two graph composition selection rules. For example, it pairs edges *e* and *f*, because *e* composed with any other edge results in a graph with three or four terminals, but *e* composed with *f* has only two terminals. On iteration 1 of the **while** loop of Fig. 4.12, the procedure Find_best_pair_of_graphs_to_merge selects edges *e* and *f* because they form the only pair that results in a two-terminal graph when composed; all other pairs compose to form three-terminal graphs. This pair of graphs is composed by procedure Compose_graph_$G_3$, with the resulting graph shown in Fig. 4.13(b). The same figure shows one of the covers, (3,7), (*e,e*), (4,8), (*f,f*), (3,9), resulting from the procedure Concatenate_CTC of Fig. 4.12. The procedure Find_best_graph_to_-merge_with pairs the resulting graph with edge *d*, and the graph count is decremented. The graph is composed by iterations 2 and 3 of the **while** loop of Fig. 4.12, as shown in Figs. 4.5(c) and (d). Figure 4.13(c) shows the 2 three-terminal dual graphs resulting from iteration 4 of the aforementioned **while** loop along, with a cover

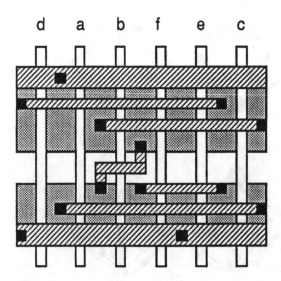

**Fig. 4.14.** Minimum-width layout of trail cover of graph in Fig. 4.13(d).

having the four trails

    (1,6) (*a,a*), (2,5);
    (1,2), (*b,b*), (3,9);
    (2,5), (*d,d*), (4,7);
    (2,6), (*c,c*), (3,7), (*e,e*), (4,8), (*f,f*), (3,9).

The fifth iteration completes the composition of the graph shown in Fig. 4.13(d) with the single dual trail (2,6), (*c,c*), (3,7), (*e,e*), (4,8), (*f,f*), (3,9), (*b,b*), (1,6), (*a,a*), (2,5), (*d,d*), (4,7) resulting from the concatenation by Concatenate_CTC of the four trails of the cover of Fig. 4.13(c). Figure 4.14 shows the layout corresponding to the cover of Fig. 4.13(d), which is of minimum width.

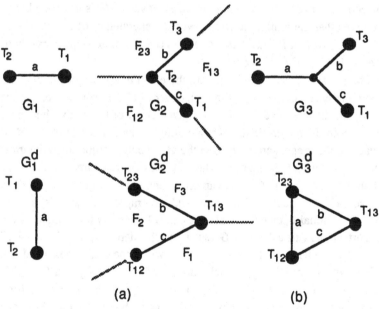

              (a)                        (b)

**Fig. 4.15.** Illustration of terminals and faces involved in dual graph composition.

## 4.2.3   Optimality

In this section, we prove that *P-TrailTrace* leads to a minimum-width layout for dual planar graphs.

**Theorem 4.2:** *P-TrailTrace* solves the W layout minimization problem exactly for any dual planar circuit.

**Proof:** The proof is by induction. The circuit is modeled by a TTPM $G$. The base case is that $G$ is a single-edge graph. *P-TrailTrace* assigns a primitive CTC to single-edge graphs that consists of the only two covers possible, based on exhaustive enumeration. In Fig. 4.15, graphs $G_1$ and $G_1{}^d$ are a single-edge pair. The CTC containing the only two primitive cover types is $\{(T_1,T_1/T_2,T_2), (T_1,T_2/T_2,T_1)\}$. Furthermore, each cover has a single dual trail. Therefore, the set is complete, and each of its covers is minimal. This solves the W layout problem for the base case.

Assume that *P-TrailTrace* computes a minimal CTC for all subgraphs for the first $n - 1$ composition steps of the graph $G$ with $n + 1$ edges. (A graph of $n$ edges requires $n - 1$ composition steps.) We now prove that based on this assumption, *P-TrailTrace* computes a minimal CTC for $G$ from its two components $G_1$ and $G_2$. There are two steps to combining the minimal CTCs of the components to form the minimal CTC of $G$. The first is that the covers of the components must be combined correctly to ensure that a minimum-sized cover of $G$ is generated. The second step requires that the minimum-sized cover so generated must be retained in the CTC of $G$. We show that *P-TrailTrace* performs each of these steps correctly to ensure that the minimal CTC of $G$ is computed.

**Step 1:** The covers of $G_1$ and $G_2$ are combined in every way possible. *P-TrailTrace* combines each cover of $G_1$ with each cover of $G_1$. Furthermore, there is only one way to concatenate the dual trails of a given cover of $G_1$ and with a cover of $G_2$. If one or both of the terminals corresponding to the endpoints of two dual trails, one in each of the graphs, are merged in the graph composition, then the trails are concatenated; otherwise, they are not. When $G_1$ and $G_2$ are composed, there can be no more than one dual trail in $G_2$ that can concatenate with a given dual trail in $G_1$. This is ensured by the duality of the trails and the graphs. If two edges in $G$ are incident with a terminal $T$, then the dual edges in $G^d$ can only be incident with a terminal that corresponds to a face of $G$ that is not bordered by $T$. Figure 4.15 shows edges $b$ and $c$ incident with terminal $T_2$ in $G_2$, where the dual edges $b$ and $c$ in $G_2{}^d$ are incident with terminal $T_{13}$, which corresponds to face $F_{13}$ in $G_2$. This face is bordered by terminals $T_1$ and $T_3$, but not by $T_2$. In graph composition involving $T$, only a terminal corresponding to a face of $G$ that is bordered by $T$ can be the dual terminal involved in concatenation with trails of the other graph. Therefore, only one such dual trail exists in $G$ and $G^d$. In Fig. 4.15, edges $b$ and $c$ are incident with $T_2$ in $G_2$. Only $T_{12}$ or $T_{23}$ can be the dual terminal of a trail in $G_2{}^d$ that can concatenate with $G_1$ and $G_1{}^d$.

**Step 2:** The resulting minimum-sized covers are included in the minimal CTC of $G$. The algorithm only discards a new cover if another cover of the same type, but smaller in size, already exists in the CTC. Moreover, it discards a cover already in the CTC if a new cover is generated that is of the same type, but smaller than it in size. Since neither of these situations can result in the loss of a minimum-sized cover, we conclude that *P-TrailTrace* retains a minimum-sized cover of each possible type for $G$.

Therefore, we have proved that the minimum-sized CTC is computed for $G$, based on the inductive hypothesis. Any minimum-sized cover in the CTC of $G$ leads directly to a minimum-width layout of the planar circuit. Consequently, *P-TrailTrace* solves the W layout problem for any dual planar CMOS circuit. $\Box$

## 4.2.4 Time Complexity

The time complexity of *P-TrailTrace* as a function of the height $h$ of the circuit graph is $O(h^4 2^{2h^2})$. The **while** loop of Fig. 4.12 is executed $O(E)$ times. The operation in this **while** loop whose complexity dominates all others is Concatenate_CTC with $O(|CTC|^2 Tmax) = O(2^{2E} Tmax)$ time; therefore, the loop has complexity $O(E2^{2E} Tmax)$. Since the complexity of this **while** loop dominates the algorithm, the overall complexity of *P-TrailTrace* is $O(E|CTC|^2 Tmax) = O(E2^{2E} Tmax)$. The size of $E$ and $Tmax$ are both $O(h^2)$, hence the complexity of the algorithm can be expressed solely in terms of $h$ as $O(h^4 2^{2h^2})$.

However, we are interested in applying *P-TrailTrace* to circuits of practical size, not to circuits of arbitrary complexity. We have observed that all nonseries-parallel circuits with $h \leq 4$ can be decomposed into subgraph having no more than three terminals. Therefore, for these circuits, the maximum size of a CTC at any stage in the composition process cannot exceed 429. This allows us to compute the maximum number of operations required to process any of the practical circuits. Therefore, for practical circuits, worst-case time is

$$c \ E \ |CTC|^2 \ Tmax \ \approx \ 9,000,000c$$

Even with $c = 1000$, running *P-TrailTrace* on a computer which can execute an average operation in 100 ns, the worst-case execution time is 900 seconds or 15 minutes. This is a reasonable limit on time and strongly suggests that *P-TrailTrace* is sufficient for practical circuits. In the next section, we show that *P-TrailTrace* can actually generate layouts for any practical-sized circuits in less than 10 seconds.

## 4.3 COMPLETE STUDY OF PRACTICAL PLANAR CELLS

We present the results of a complete computer-aided analysis of the class of all practical-sized circuits using *P-TrailTrace*. First, we define this class of circuits. Next we present data on the number of terminals needed to compose any circuit of this class, the number of covers generated in the composition process, and run-time statistics for *P-TrailTrace* on this class of circuits.

We have identified all distinct nonseries-parallel TTPM of height $h \leq 4$ by a systematic manual search. We started with the bridge circuit of Fig. 4.16, having height three, which every nonseries-parallel graph must contain as an embedded graph [Du65]. We systematically added vertices and edges to the bridge graph and its dual, such that the resulting graph was planar and its height was limited to four or less. There are 144 such circuits, which constitute the set of all TTPMs we consider to be of practical size. We have studied this set both to learn about its characteristics and to understand better the performance of *P-TrailTrace*. This study includes producing layouts for all the practical circuits using our programmed implementation of *P-TrailTrace*.

We initially hypothesized that in composing the 144 practical circuits, none would require any of its component subgraphs to have more than three or four terminals. Our experiments show that, in fact, at most three terminals are needed. These data demonstrate that the graph composition algorithm Compose_Graph indeed finds an optimal sequence of compositions, resulting in the minimum number of terminals in the component graphs over the whole composition process. Compose_Graph, by its pair-selection rules, combines the components that result in three-terminal graphs only if there is no pair that results in two terminals. Those that the algorithm composes with a maximum of three terminals cannot be composed with only two terminals. Hence, all practical-sized graphs are composed with the minimum number of terminals by *P-TrailTrace*.

We present data in Fig. 4.17 showing the optimal cover-size distribution and the final CTC size for all practical-sized nonseries-parallel planar circuits. It is not surprising to see that 89% of these circuits require two trails to cover them, since the bridge circuit, which each of these circuits contains, requires two trails to cover it. We prove this fact below. However, it is interesting that 10 circuits included in Fig. 4.17 have d-euler trails.

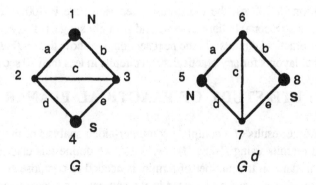

**Fig. 4.16.** The bridge graph and its dual.

| Number of dual trails in optimal cover | Number of circuits | Percent of circuits |
|:---:|:---:|:---:|
| 1 | 10 | 7 |
| 2 | 128 | 89 |
| 3 | 6 | 4 |

**Fig. 4.17.** Optimal cover-size distribution for all practical-sized TTPMs.

**Theorem 4.3:** The bridge circuit and its dual require at least two d-trails to cover them.

**Proof:** Figure 4.16 shows a bridge graph $G$ and its dual $G^d$, which is also a bridge graph. A graph is series-parallel if and only if no embedded graph is a bridge [Du65]; therefore, the bridge characterizes nonseries-parallel graphs. Each of the graphs in Fig. 4.16 has only one pair of vertices of odd degree, so each graph has an euler trail. In order for them to have a d-euler trail, it is necessary that the d-euler trail start and end at the odd vertices in each graph. Vertices 2 and 3 in $G$ are odd, as are 6 and 7 in $G^d$. In $G$, if one end of the trail is at vertex 2, then the other end must be at vertex 3. The edges incident upon vertex 2 in $G$ are $a$, $c$ and $d$. If $a$ is the first edge, then vertex 6 must be the endpoint of the trail in $G^d$, since it is the odd vertex adjacent to $a$. Once this choice of starting edge and endpoints in $G$ and $G^d$ is made, the next edge must be incident upon vertices 1 and 5. But no edge satisfies this requirement. Therefore, edge $a$ cannot be a first edge in a d-euler trail. A similar argument for the other edges shows that none of them can be a first edge in the d-euler trail. Since every d-euler trail must have a first edge, we have proved that the bridge circuit and its dual do not have a d-euler trail. It is easy to see that these graphs can be covered by two d-trails. For example, the d-trails $(2,5)$, $(a,a)$, $(1,6)$, $(b,b)$, $(3,8)$, $(e,e)$, $(4,7)$, $(d,d)$ and $(2,6)$, $(c,c)$, $(3,7)$ constitute a minimum cover. $\square$

Figure 4.18 shows the distribution of the number of covers in the CTCs of all practical-sized nonseries-parallel circuits. The upper bound on the CTC size is 29. In fact, we found that the maximum number of covers at any composition step for any of these circuits is 80, with the average number about 25. This confirms our hypothesis that the number of covers is small for such circuits.

| Number of covers in final CTC | Number of circuits | Percent of circuits |
|:---:|:---:|:---:|
| 1-9 | 4 | 3 |
| 10-19 | 66 | 46 |
| 20-29 | 74 | 51 |

**Fig. 4.18.** Final CTC size distribution for all practical-sized nonseries-parallel TTPMs.

| Execution time (s) | Number of circuits | Percent of circuits |
|:---:|:---:|:---:|
| 0-1 | 1 | 1 |
| 1-2 | 10 | 14 |
| 2-3 | 25 | 35 |
| 3-4 | 20 | 28 |
| 4-5 | 8 | 11 |
| 5-6 | 5 | 7 |
| 6-7 | 2 | 3 |
| 7-8 | 1 | 1 |

**Fig. 4.19.** Actual execution-time distribution for all practical-sized planar circuits.

The distribution of the execution times required for *P-TrailTrace* to generate a layout for each of the planar circuits is given in Fig. 4.19; again the computer used is a MicroVAX II. Note that the worst-case run time is less than 8 s and that the average is about 3 s. This clearly shows that *P-TrailTrace* executes in a reasonable amount of time for all these circuits. Figure 4.20 shows the ratio between the maximum number of cover concatenations required to cover a graph by exhaustively retaining all covers and the number of covers using cover types to eliminate all covers of a given type but the smallest one. By using cover types and only retaining one cover of each type, the number of cover concatenation operations is reduced on average by more than 50%.

| Percent of maximum concatenations done | Number of circuits | Percent of circuits |
|:---:|:---:|:---:|
| 100 | 8 | 5 |
| 80-99 | 10 | 7 |
| 60-79 | 36 | 25 |
| 40-59 | 24 | 17 |
| 20-39 | 45 | 32 |
| 1-19 | 21 | 14 |

**Fig. 4.20.** Efficiency of graph covering in *P-TrailTrace* on all practical-sized planar circuits.

# CHAPTER V
# SINGLE CELL WIDTH AND HEIGHT
# MINIMIZATION

The three cell layout algorithms presented in Chapters 3 and 4 minimize the width of a functional cell but do not explicitly consider the cell's height. This chapter addresses both width and height minimization for the case of dual series-parallel CMOS cells. We extend the theory of trail covering to deal with height, and present an algorithm *HR-TrailTrace*, which is a generalization of *R-TrailTrace* presented in Chapter 3. This algorithm generates for a given CMOS circuit modeled by a TTSPM $M$ and its dual a layout of minimum height with respect to all layouts of minimum width for any reordering $M'$ of $M$, thus solving the WHR layout problem for series-parallel circuits.

The chapter is organized as follows. In Section 5.1, the cell layout style is presented and the cell height problem is defined. In Section 5.2, we describe extensions to the series-parallel cell theory to address cell height. The algorithm *HR-TrailTrace* is presented in Section 5.3, and its optimality is proved. The time complexity of the algorithm is analyzed, both from a worst-case point of view and from experimental data collected from an implementation of the algorithm. In Section 5.4, the results of a complete computer study of the height of all practical size cells are presented with a comparison to other layout methods. Section 5.5 discusses how this theory can be extended to include dual planar circuits.

## 5.1 LAYOUT PROBLEM

The three algorithms presented in Chapters 3 and 4, *TrailTrace*, *R-TrailTrace* and *P-TrailTrace*, produce minimum-width cells, but ignore the cells' height. Such algorithms are suitable to design environments where the height of the overall row of cells is predetermined by technology considerations [Uv81]. For instance, if the custom cells are being mixed with cells from a standard cell library, and if the height of these library cells is set to accommodate the tallest cell in the library, then the height of the row is determined by the height of these library cells and cannot be decreased even if all the custom cells are shorter than this height. However, where the row is composed solely of custom cells, the row height is determined by the height of its tallest custom cell. In this case, the cell height is very important and must be minimized to achieve an efficient layout. A quantitative study of a broad class of circuits presented below demonstrates that area differences of more than 80% can result by minimizing height as a secondary optimization goal to the primary width minimization goal.

### 5.1.1  Constraints and Assumptions

While the width minimization problem is well-defined in prior work [Uv81], the height minimization problem has not been so clearly defined. Therefore, we first specify all assumptions about our layout style, most of which are implicit in the layout systems described in the literature. These assumptions clarify and constrain the problem in ways that yield three benefits: optimality arguments are facilitated, the search for optimal layouts is speeded up considerably, and the resulting layouts are area-efficient. It should be carefully noted that this layout style is not the only one that is suitable for our layout method. Our layout algorithms can be applied to other layout styles that use less diffusion and result in faster circuits; one example is the two layer metal style shown in [Ch91]. The style we consider here is simply one example and was chose to facilitate area comparisons with previous layout methods.

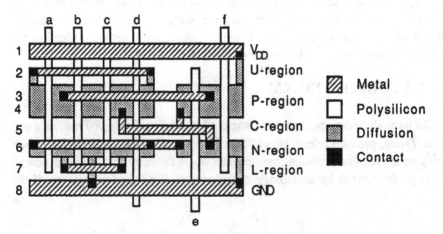

**Fig. 5.1.** An illustration of our single-cell layout style.

The single-cell layout style under consideration is illustrated in Fig. 5.1. Five routing regions are defined in the cell: the N- and P-regions corresponding to the areas of nMOS and pMOS transistors, the Upper (U) region above the P-region, and Lower (L) region below the N-region, and the Center (C) region between the N- and P-regions. Figure 5.2 lists each of the constraints and assumptions characterizing the layout style of Fig. 5.1. Metal is the primary method of horizontal interconnection; diffusion is used in this direction only to connect drain and source terminals, according to assumption 5. Vertical connections are made primarily with polysilicon, according to assumption 6. However, assumption 7 allows a limited amount of diffusion to be used to make vertical connections from transistor drain and source terminals to horizontal metal lines. It also allows vertical metal wires to be used only to interconnect the pullup and pulldown subcircuits of a functional cell.

We next illustrate several of the assumptions of the layout style of Fig. 5.1. Assumptions 1 through 3 are explicitly stated by Uehara and vanCleemput [Uv81]; see Fig. 2.1. Assumptions 4, 5 and 6 are implied by illustrations in [Uv81]. Assumption 7 is illustrated in Fig. 5.1, where diffusion is used on the right of column $f$, routing $V_{DD}$ in row 1 to the pMOS transistors in row 3. It is also used between columns $c$ and $d$ between rows 6 and 7 to connect the drain/source terminal to the horizontal metal line in row 7. Assumption 7 is also illustrated by the metal connection between columns $c$ and $f$ between rows 4 and 6 of Fig. 5.1. According to assumption 11, intracell routing of the pullup subcircuit is done in the P-region, as in row 3 of Fig. 5.1 or, if this region is full, in the U-region, as shown in row 2 of the figure. Likewise the N- and L-regions are used for intracell routing of the pulldown subcircuit, as assumption 11 states. This is illustrated in rows 6 and 7 in Fig. 5.1. The bottoms of the n-channel transistors and tops of the p-channel transistors are aligned to the C-region, as determined by assumption 15, and also shown in Fig. 5.1.

## 5.1.2 Comparison to Other Assumptions

Most of these assumptions are used by other layout methods published in the literature. Few of these layout papers explicitly and exhaustively list the constraints imposed on their layouts; in many cases, it is necessary to surmise them by studying the examples given in the papers. An assumption is attributed to a particular method if its layouts exhibit the assumption almost universally.

Although some previous layout methods adopt assumption 7, it has not been as universally employed as the others; therefore, we now discuss it in detail. This assumption allows vertical diffusion to be used to connect source or drain terminals

1.  Only static dual series-parallel CMOS circuits are used.
2.  The transistors are placed in two parallel horizontal rows, the upper row for the pMOS and the bottom row for the nMOS transistors. This allows single p- and/or n-wells to be used. The channels of the transistors are oriented so that their length dimension is in the horizontal direction.
3.  The gate terminals of complementary transistors are vertically aligned to share the same column.
4.  Three interconnection layers are used: diffusion, polysilicon and one level of metal.
5.  All horizontal connections are made in metal. One exception is that horizontal connections between adjacent drain and/or source terminals of neighboring transistors are made in diffusion.
6.  All vertical connections between horizontal metal and gate terminals and between complementary gates of the pullup and pulldown subcircuits of a given functional cell are made in polysilicon.
7.  All vertical connections between source/drain terminals and metal are in diffusion. The only use of metal for vertical connections is between the pullup and pulldown subcircuits of a functional cell.
8.  No jogs are used in any layer except in connecting the two subcircuits of a given functional cell.
9.  Layer changes to jump to another track are not used.
10. Only uniform-size transistors are used in the pullup and pulldown subcircuits.
11. All intracell routing of the pullup subcircuit is done in the P- or U-regions.
12. All intracell routing of the pulldown subcircuit is done in the N- or L-regions.
13. The pullup and pulldown subcircuits of a cell are interconnected in the C-region between the side boundaries of the cell.
14. $V_{DD}$ and GND are routed through the U-and L-regions, respectively.
15. All p- and n-channel transistors are aligned to the C-region.
16. A single contact from metal to diffusion or polysilicon is used.

**Fig. 5.2.** Assumptions characterizing the layout style of Fig. 5.1.

to horizontal metal lines for intracell routing. TOPOLOGIZER uses such a scheme for both intra- and intercell routing [KW85]. SOLO uses vertical diffusion for intracell connections [BA88]. The methods of Madsen, Lefebvre and Chan, and that of Poirier use vertical diffusion to route $V_{DD}$ and GND [Ma89, LC89, Po89]. These systems suggest that a careful use of diffusion for intracell and intercell interconnection is acceptable in this context. The concerns about using diffusion are that it has a high resistance and high capacitance, which can increase delay and reduce circuit performance. For instance, a 100 μm x 4 μm diffusion wire has about 30 times more capacitance than a polysilicon wire of equal size, with half the resistance, based on a representative 3-5 μm CMOS process described in [WE85]. Therefore, the RC time constant for the diffusion wire is about 17 times larger than for the polysilicon wire. Clearly, long diffusion lines have a more significant affect on circuit speed than polysilicon or metal in cases where capacitance is critical. Therefore, wherever diffusion is used for interconnection in our style, we reduce its length as much as possible. Although not shown in Fig. 5.1, six of the seven vertical diffusion lines used in that layout can be replaced in a postprocessing step by metal wires without increasing the layout area.

We will show that cells in our layout style are as small as or even smaller than cells whose styles are not as constrained. This is because our method does a more extensive search over a reasonable set of constraints. Other methods often do a less extensive search with a fairly similar set of constraints.

With our layout assumptions (Fig. 5.2), we can now define the corresponding cell layout minimization problem formally. The cell height $H$ is given by the following equation

$$H = U + P + C + N + L$$

where $U, P, C, N$ and $L$ are the number of rows in the U-, P-, C-, N- and L-regions, respectively, as shown in Fig. 5.1. The C-region is always one row high in our style, and the P- and N-regions are fixed for a given circuit by the dimensions of the transistors; they do not vary between different layouts of the same circuit. The cell width $W$ is given by the equation

$$W = T + G + 1$$

where $T$ is the number of transistors in either the pullup or pulldown circuit, and $G$ is the number of diffusion gaps required for a given layout [Uv81]. Based on the above formulas for cell width and height, we define the *WHR layout minimization problem* as follows: given a dual series-parallel circuit with circuit graph $M$, find a layout in our single-cell layout style with minimum $H$ from among all layouts of minimum $W$, for any reordering $M'$ of $M$.

## 5.2 EXTENSION OF SERIES-PARALLEL CELL THEORY

This section discusses the main extensions to the series-parallel cell theory of Chapter 3 to deal with cell height. First, we generalize the dual-trail-cover equivalence relation of *R-TrailTrace* that determines which covers are retained or discarded in the process of composing and covering the circuit graph $M$ and its dual. *R-TrailTrace* includes a cover in the CTC if and only if either its cover type is unique, or no cover of the same type has fewer trails. The result of using this equivalence relation is that the final CTC is a UMCTC as defined in Chapter 3, i.e., it contains only one cover of minimum size for each cover type. This equivalence relation is sufficient for finding the minimum-width cell layout. However, it does not take the cell height into account. The algorithm *HR-TrailTrace* we present in this chapter considers two trail covers of $M$ equivalent if and only if they consist of precisely the same trails. This ensures that the trail covering procedure of the algorithm retains every cover that possibly could lead to a layout of minimum height and width. The resulting CTC we define formally as follows. A *plenary minimum complete trail cover* (PMCTC) of $M$ is a CTC that consists exclusively of every minimum-sized cover of each possible cover type of $M$. A PMCTC of $M$ is a superset of a UMCTC of $M$. We have shown above for *R-TrailTrace* that the maximum size of a UMCTC of $M$ is 42. In the worst case, the size of a PMCTC of $M$ is $O(E!)$, where $E$ is the number of edges of $M$. This is excessively large even for small $E$. Despite this upper bound, we shall show below that the PMCTC concept used in *HR-TrailTrace* leads to reasonable execution times for all static functional cells of practical size.

A second difference between the two algorithms is in the procedure for generating a layout from a trail cover. This involves ordering the trails in a horizontal row and orienting each trail so that a given end of the trail points in either the left or right direction. *R-TrailTrace* does not consider these various permutations or orientations of the trails, since they do not affect the optimal width of the cell. However, they do affect the cell height significantly. Therefore, *HR-TrailTrace* generates all permutations and orientations of the trails of a cover to find the optimal height. Our experiments show that this is computationally feasible for practical-sized circuits.

A third difference involves the case of closed dual trails. A *closed dual trail* is a dual trail in $M$ and $M^d$, where the start and end point of each trail is the same [Bo79]. *R-TrailTrace* ignores closed dual trails, since they do not affect cell width. The circuit in Fig. 5.3, which appeared earlier in Fig. 3.10, has a single closed dual trail cover. In order to create a layout from this closed dual trail, it must be broken at some point to create the horizontal diffusion lines. The points at which the closed dual trail is broken can affect the layout height. The closed dual trail of Fig. 5.3(b), broken between edges $a$ and $c$, is the trail $abefhgdc$ with the resulting layout shown in Fig. 3.11. Note that the cell height is eight. The same closed trail, broken between $h$ and $f$, is the dual trail $hgdcabef$, whose layout is shown in Fig. 5.4, with a cell height of seven. *HR-TrailTrace* detects every closed dual trail in the trail covering procedure, so as not to retain covers that differ only in the position in which they break the trail. In producing a layout from such a cover, *HR-TrailTrace* breaks a closed dual trail in every place that can affect the cell height.

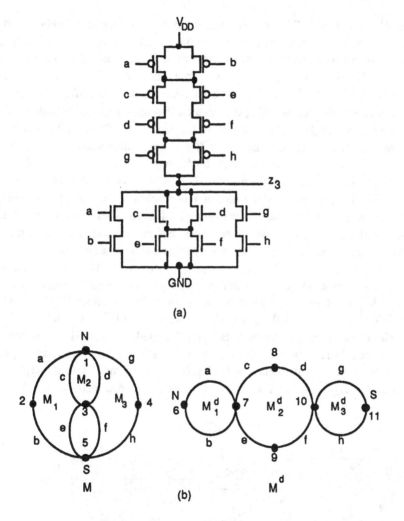

(a)

(b)

**Fig. 5.3.** The (a) circuit and (b) graph model of Fig. 3.10 illustrating a closed dual trail.

## 5.3 HR-TRAILTRACE ALGORITHM

We now present the *HR-TrailTrace* algorithm, prove the optimality of its layouts, and analyze its time complexity. We also illustrate the algorithm with a detailed design example.

### 5.3.1    Description

The algorithm *HR-TrailTrace* presented here solves the WHR layout minimization problem exactly for a series-parallel circuit. It produces a cell of least height from among all those of minimum width for any reordering of the circuit. The algorithm is described in Fig. 5.5 and is a generalization of the *R-TrailTrace* algorithm described in Chapter 3.    A brief description of the *HR-TrailTrace* algorithm is given below highlighting the salient differences from *R-TrailTrace*.

First, we give a high-level description of the algorithm. The circuit graph $M$ is entered, and *HR-TrailTrace* constructs from $M$ the composition tree $T$. It traverses $T$ in bottom-up fashion, composing a subgraph $M_i$ represented by each node $T_i$ of $T$ and computing PMCTC($M'_i$) from the PMCTCs of the children of node $T_i$ . When the tree traversal is complete and the root of $T$ is reached, PMCTC($M'$)  containing all minimum-sized covers of each cover type of all reorderings $M'$ of $M$  has been computed. This set is pruned of all covers exceeding the minimum number of trails. From the pruned set, all layouts of potentially minimum height are generated. The minimum numbers of rows $U$ and $L$  required to route the U- and L- regions, respectively, are measured for each such layout and are represented by a *height vector* $(U, L)$.    A height vector $(U_i, L_i)$ *dominates*  another height vector $(U_j, L_j)$ if and only if  $(U_i < U_j$ and $L_i < L_j)$, $(U_i = U_j$ and $L_i < L_j)$ or $(U_i < U_j$ and $L_i = L_j)$. Each nondominated height vector with its layout is included in a set HV, from which *HR-TrailTrace* selects a minimum-height layout as its final result.  For example, the height vector for the layout in Fig. 5.4 is $(U_1, L_1) = (3, 3)$, and that of the layout in Fig. 3.11 is $(U_2, L_2) = (4, 3)$. The height vector $(U_1, L_1)$  dominates $(U_2, L_2)$.

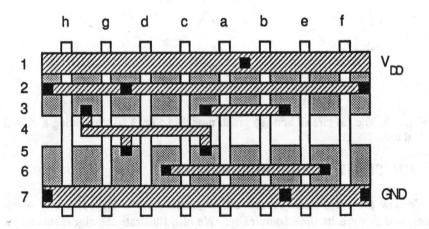

**Fig. 5.4.** Layout of optimal height and width for the circuit  in Fig. 5.3.

```
procedure Generate_min_height_layout(2)(cell, Min_covers, HV);
begin
for each TC in Min_covers do
        for each break of each closed dual trail in TC do
            for each permutation of trails in TC do
                for each reflection of the trails of TC do
1:                  for each permutations of signals of isomorphic subgraphs do
                        begin
                        U:= 0;  L:= 0;
                        for each column of layout do
                            begin
                            Count number of wire_crossings in U- and L-regions at column;
                            if (wire_crossings_in_U > U) then U:= wire_crossings_in_U;
                            if (wire_crossings_in_L > L) then   L:= wire_crossings_in_L;
                            end;
                        if no (U_cell,k, L_cell,k ) in HV_cell dominates or equals (U,L) then
                            Save (U,L) and its layout in HV_cell;
                        end;
end; { Generate_min_height_layout(2) }
```

```
function Ordered_ children_CTC(Node_type, Child_CTC, Child_order): CTC_type;
begin  Ordered_ children_CTC := Child_CTC[Child_order[1]];
for  i := 2 to Child_count of Node do
    begin
    Temp_ children_CTC := empty;
    for each trail cover in Ordered_children_CTC do
        for each trail cover in Child_CTC[Child_order[i]] do
            begin
            Concatenate both trail covers according to Node_type operation;
2:          if result TC type according to HR-TrailTrace cover equivalence relation does
            not exist in Children_CTC then Add trail cover to Temp_children_CTC
            else if result trail cover has fewer internal trails than does TC with same TC
            type in Temp_ children_CTC  then  Subst. result cover for larger cover;
            end;
        Ordered_ children_CTC := Temp_ children_CTC;
    end;
end; {Ordered_ children_CTC}
```

**Fig. 5.5.** The *HR-TrailTrace* layout algorithm.

The subprograms mentioned above will now be described more fully. The first is Node_CTC. As in *R-TrailTrace*, this recursive function traverses $T$ in postorder. Each subtree of $T$ represents a subgraph of $M$, with the leaves corresponding to the

---

```
function  Node_CTC(Node):  CTC_type;
begin  Node_CTC := empty;
if Node is a leaf  then  Node_CTC := Edge_CTC    {Edge_CTC consists of the two covers
                                                          of an edge}
else  begin
      for  i := 1 to Child_count of Node  do  Child_CTC[i] := Node_CTC(Child[i]);
      for each distinct Child_order do
            begin
            Temp_node_CTC := Ordered_children_CTC(Node_type, Child_CTC, Child_order);
3:          Update Node_CTC with new or better trail covers from Temp_node_CTC based
            on cover equivalence relation;
            end;
      end;
end;  {Node_CTC}

procedure  HR-TrailTrace(M);
begin
Input(M);
Construct composition tree T from M;
Graph_CTC := Node_CTC(T);
Min_covers:= Retain for layout all covers of minimum size from Graph_CTC;
Generate_min_height_layout(Min_covers, HV ) and select minimum height layout from
HV;
end.  {HR-TrailTrace}
```

---

**Fig. 5.5. (continued)**   The *HR-TrailTrace* layout algorithm.

---

edges of $M$, and the internal nodes of $T$ corresponding to the */+ or +/* operations
on series-parallel graphs. When a leaf is visited, the CTC for a single-edge graph is
the cover associated with it. As each internal node of $T$ is subsequently visited, the
CTCs of all the node's children are combined according to the composition operation
of the node in bottom-up fashion. As in *R-TrailTrace*, the order of the children of a
node is permuted, and for each such permutation the composition operation is applied
to the children of the node by calling Ordered_children_CTC with the composition
operation, the children of the node, and a permutation of their order.
Ordered_children_CTC returns the CTC for the node corresponding to the
permutation of the children. This CTC is combined with those retained from any
previous permutation of the children. At each stage in this process, after all distinct
permutations of the children of a node of $T$ have been performed, the resulting CTC
of this node contains all the potentially optimal covers for the subgraph of $M$ to
which the node corresponds. The process ends when the root node of $T$ is visited, and
the corresponding CTC of the root is the optimal CTC of $M$. As in *R-TrailTrace*,
the order of the children of a node is permuted in all distinct ways, and the
composition operation is applied for each such permutation. As described above, a

cover equivalence relation different from that of *R-TrailTrace* is applied to the resulting covers in statements 2 and 3 of Fig. 5.5, so that the cell height is accounted for, as well as the width.

The list of covers called Min_covers results from pruning Graph_CTC of all covers that have more than the minimum number of trails. Trail cover type is ignored in this pruning; cover size is the only consideration. The procedure Generate_min_height_layout produces layouts for each cover in Min_covers. Each closed dual trail is broken, the trails in the cover are placed by permuting their order, and each trail is oriented in its position in one of the two possible ways. For each such layout, the maximum number of horizontal interconnections is measured for the U- and L-regions. A layout is retained if its $(U, L)$ height vector is not dominated by any other. This procedure generates HV, the set of all nondominated height layouts of minimal width. A layout of minimum height among all layouts of minimum width is selected from HV.

## 5.3.2  Design Example

We illustrate the *HR-TrailTrace* algorithm by an example. Figure 5.6(a) shows a transistor circuit implementing the function $z = \neg(*(+a(*bc))(+de(*(+fg)(+hi))))$, along with the corresponding graph model in Fig. 5.6(b), where $M$ and $M^d$ represent the pulldown and pullup subcircuits, respectively. The algorithm begins by building the composition tree $T$ for the circuit; see Fig. 5.7. The function Node_CTC of *HR-TrailTrace* traverses $T$ in the same bottom-up manner as *R-TrailTrace*. It first composes edges $g$ and $f$ in parallel in $M$ and the dual edges are composed in series in the dual graph, according to the composition subexpression $(+/*gf)$. At the same time, the primitive covers of edges $f$ and $g$ are concatenated together. This is followed by composing edges $i$ and $h$ in parallel in $M$, and so forth. We look in detail at the last composition operation at the root of $T$, where the graphs $M_1$ with edges $d, e, f, g, h$ and $i$, and $M_2$ with edges $a, b$ an $c$ in Fig. 5.6(b) are composed by the function Ordered_children_CTC in Fig. 5.5 along with their dual graphs according to the */+ operation. Let $ctc_i$ be a subset of the PMCTC for $M_i$.

$$ctc_1 = \{ tc_1, tc_2 \}$$
$$ctc_2 = \{ tc_3, tc_4, tc_5, tc_6 \}$$

Now the trail covers of $ctc_1$ are

$$tc_1 = \{ (3,11), (a,a), (5,12), (c,c), (4,6), (b,b), (3,12) \}$$
$$tc_2 = \{ (3,11), (a,a), (5,12); (5,6), (c,c), (4,12), (b,b), (3,6) \}$$

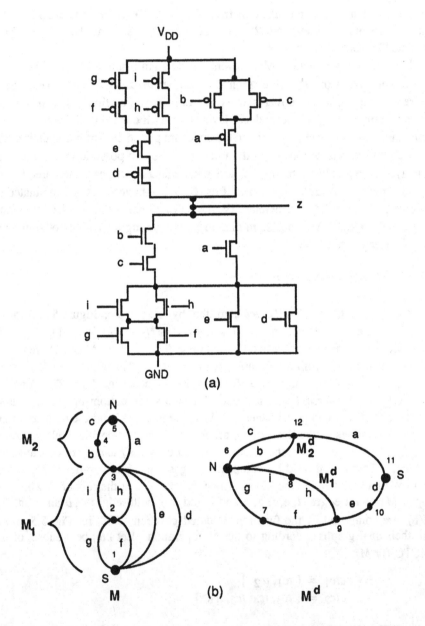

**Fig. 5.6.** (a) Transistor circuit and (b) graph model implementing the function
$z = \neg(*(+a(*bc))(+de(*(+fg)(+hi))))$.

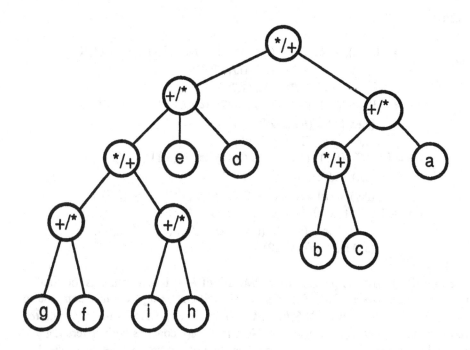

**Fig. 5.7.** Composition tree T for the circuit of Fig. 5.6.

The set of trail covers for $ctc_2$ consists of

$tc_3 = \{$ (3,11), (d,d), (1,10), (e,e), (3,9), (h,h), (2,8), (i,i), (3,6);
       (1,6), (g,g), (2,7), (f,f), (1,9) $\}$
$tc_4 = \{$ (3,11), (d,d), (1,10), (e,e), (3,9), (h,h), (2,8), (i,i), (3,6);
       (2,6), (g,g), (1,7), (f,f), (2,9) $\}$
$tc_5 = \{$ (3,11), (d,d), (1,10), (e,e), (3,9);
       (2,9) (h,h), (3,8), (i,i), (2,6), (g,g), (2,7), (f,f), (2,9) $\}$
$tc_6 = \{$ (1,11), (d,d), (3,10), (e,e), (1,9), (f,f), (2,7), (g,g), (1,6);
       (3,6), (i,i), (2,8), (h,h), (3,9) $\}$.

The covers $ctc_1$ and $ctc_2$ are combined pairwise, resulting in $ctc_0$, called Graph_CTC in Fig. 5.5, with four covers of $M$.

$ctc_0 = \{ tc_7, tc_8, tc_9, tc_{10} \}$

where

$$tc_7 = \{(3,12),\ (b,b),\ (4,6),\ (c,c),\ (5,12),\ (a,a),\ (3,11),\ (d,d),\ (1,10),$$
$$(e,e),\ (3,9),\ (h,h),\ (2,8),\ (i,i),\ (3,6);$$
$$(1,6),\ (g,g),\ (2,7),\ (f,f),\ (1,9)\ \}$$

$$tc_8 = \{(3,12),\ (b,b)\ ,\ (4,6),\ (c,c),\ (5,12),\ (a,a),\ (3,11),\ (d,d),\ (1,10),$$
$$(e,e),\ (3,9),\ (h,h),\ (2,8),\ (i,i),\ (3,6);$$
$$(2,6),\ (g,g),\ (1,7),\ (f,f),\ (2,9)\ \}$$

$$tc_9 = \{\ (3,12),\ (b,b)\ ,\ (4,6),\ (c,c),\ (5,12),\ (a,a),\ (3,11),\ (d,d),\ (1,10),$$
$$(e,e),\ (3,9);$$
$$(2,9),\ (h,h),\ (3,8),\ (i,i),\ (2,6),\ (g,g),\ (2,7),\ (f,f),\ (2,9)\ \}$$

$$tc_{10} = \{\ (1,11),\ (d,d),\ (3,10),\ (e,e),\ (1,9),\ (f,f),\ (2,7),\ (g,g),\ (1,6);$$
$$(5,6),\ (c,c),\ (4,12),\ (b,b),\ (3,6),\ (i,i),\ (2,8),\ (h,h),\ (3,9);$$
$$(3,11),\ (a,a),\ (5,12)\ \}.$$

Each resulting cover is placed in the PMCTC of $M$ if it is as small as, or smaller than, any other cover of $M$ of its type. The trail covers $tc_2$ and $tc_4$ combine to form $tc_7$, which is of type (N,S)/(I,I). Cover $tc_1$ concatenates with $tc_3$ and $tc_4$ resulting in $tc_8$ and $tc_9$, respectively. Note that $ctc_0$ contains both covers $tc_8$ and $tc_9$ of the same (I,I)/(I,I) cover type and these two covers are also the same size, i.e., two trails each. This is because $ctc_0$ is a PMCTC and not a UMCTC, as computed by *R-TrailTrace* where only cell width is of concern. Any cover of a given type and minimum size is sufficient to assure a minimum-width cell, but both covers are included because at this point it is not clear which cover leads to an minimum-height solution. The next step is to prune Graph_CTC of all covers of minimum size regardless of cover type; the resulting set is called Min_covers in Fig. 5.5. The covers $tc_7$, $tc_8$ and $tc_9$, have two trails each, whereas $tc_{10}$ from the combination of $tc_2$ and $tc_6$ has three trails, and has a cover type different from the other covers. Therefore, $tc_{10}$ is eliminated from Graph_CTC resulting in the following set:

Min_covers = $\{\ tc_7,\ tc_8,\ tc_9\}$

Layouts are generated by Generate_min_height_layout for each of these covers by breaking closed trails in all possible ways, and permuting and reflecting the trails. The cover $tc_9$ has the closed trail $(2, 9)\ (h, h),\ (3, 8),\ (i, i),\ (2, 6),\ (g, g),\ (2, 7),$ $(f, f),\ (2, 9)$; it is closed because both its endpoint pairs are identical. In generating its layouts, the trail is broken at each of its four vertices to create an open trail for subsequent placement. Cover $tc_7$ results in the minimum-height layout. Now

$$tc_7 = \{\ t_1,\ t_2\ \}$$

where

$$t_1 = \quad (3, 12),\ (b, b)\ ,\ (4, 6),\ (c, c),\ (5, 12),\ (a, a),\ (3,11),$$
$$(d, d),\ (1,10),\ (e, e),\ (3, 9),\ (h, h),\ (2, 8),\ (i, i),\ (3, 6)$$
$$t_2 = \quad (1,6),\ (g, g),\ (2, 7),\ (f, f),\ (1, 9)$$

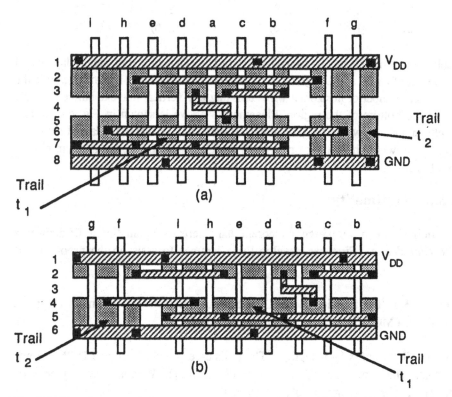

**Fig. 5.8.** Two layouts of the same cover for the circuit of Fig. 5.6(a) with different placements and orientations.

The eight permutations and reflections of these two trails result in the placement orders listed below. The symbol $t_i$ between its endpoint pairs represents the trail's placement and its orientation. For example, in row one, trail $t_1$ is placed to the left of $t_2$; trail $t_1$ is oriented such that the endpoint pair (3, 12) is placed to the left of (3, 6).

| | | | | | | | |
|---|---|---|---|---|---|---|---|
| 1. | (3, 12) | $t_1$ | (3, 6) | (1, 6) | $t_2$ | (1, 9) |
| 2. | (3, 12) | $t_1$ | (3, 6) | (1, 9) | $t_2$ | (1, 6) |
| 3. | (3, 6) | $t_1$ | (3, 12) | (1, 6) | $t_2$ | (1, 9) |
| 4. | (3, 6) | $t_1$ | (3, 12) | (1, 9) | $t_2$ | (1, 6) |
| 5. | (1, 6) | $t_2$ | (1, 9) | (3, 12) | $t_1$ | (3, 6) |
| 6. | (1, 9) | $t_2$ | (1, 6) | (3, 12) | $t_1$ | (3, 6) |
| 7. | (1, 6) | $t_2$ | (1, 9) | (3, 6) | $t_1$ | (3, 12) |
| 8. | (1, 9) | $t_2$ | (1, 6) | (3, 6) | $t_1$ | (3, 12) |

Each of these placements is scanned left to right by Generate_min_height_layout and the height of its U- and L-regions is measured. The height of a routing region is the maximum number of horizontal wires crossing any vertical line extending from the top to the bottom of the region. Placement 4 above is shown in Fig. 5.8(a), and results in a layout whose height is eight. Placement 7 results in the layout shown in Fig. 5.8(b), whose height vector is $(U, L) = (2,3)$, and whose height is the minimum for all layouts.

### 5.3.3 Optimality

In this section, we present a theorem that asserts the optimality of the algorithm *HR-TrailTrace*. First, we develop three lemmas to assist in the proof of this theorem.

**Lemma 5.1:** The function Node_CTC computes the plenary minimum complete trail cover PMCTC($M'$) for any reordering $M'$ of $M$.

**Proof:** The proof is by induction. The basis case is a single-edge TTSPM $K_1$ and its dual. The primitive CTC of $K_1$ is its PMCTC($K_1$). Without loss of generality, let $M_1,..., M_n$ be subgraphs of $M$ such that

$$M_n = (*/+ M_1,..., M_{n-1}).$$

Assume the function Node_CTC computes PMCTC($M'_1$),...,PMCTC($M'_{n-1}$). We show that it computes PMCTC($M'_n$) from PMCTC($M'_1$),...,PMCTC($M'_{n-1}$). The Node_CTC combines the $n-1$ PMCTCs by applying the */+ cover concatenation operation to them pairwise.

$$\text{PMCTC}(M_n) = \cup \ \{ \ (*/+ \ \text{PMCTC}(M'_{p(1)}),\ldots, \text{PMCTC}(M'_{p(n-1)})) \ \}$$

for each distinct permutation p of the PMCTCs of nonisomorphic subgraphs. In the application of the */+ cover concatenation operation, (*/+ PMCTC($M'_{p(i-1)}$), PMCTC($M'_{p(i)}$) ), each cover in PMCTC($M'_{p(i-1)}$) is combined pairwise with each cover in PMCTC($M'_{p(i)}$). The dual trails of each cover concatenate if their endpoint pairs are terminals of $M'_{p(i-1)}$ and $M'_{p(i)}$ that merge when the graphs are composed. Only the minimum-sized covers of each resulting cover type are kept. Such covers are necessary and sufficient to compute all minimum-sized covers of the graph. This cover composition procedure combines all minimum-sized covers of each of the subgraphs with those of its neighboring subgraphs, and does so for all distinct orderings of subgraphs. The systematic and exhaustive nature of this procedure guarantees that each minimum-sized cover of each cover type of $M_n$ is generated. $\square$

**Lemma 5.2:** Let $C$ be a routing channel with $c$ vertical columns and $r$ horizontal rows, and $W$ be a set of $n$ wires, $\{(l_1, r_1),\ldots, (l_n, r_n)\ \}$, where wire $i$ extends from column $l_i$ to column $r_i$, and $l_i < r_i$. Assume that all routing is done on a single routing layer, and that a wire $i$ can be assigned to any one row of the channel. Let $m$ be the maximum number of the wires of $W$ crossing any column of $C$. All the wires of $W$ can be routed within channel $C$ if and only if $m \le r$.

**Proof:** The lemma is proved by contradiction. The wires cannot be routed if $m > r$; therefore, the necessity condition follows. We must prove that it is sufficient that $m \le r$ for the wires of $W$ to be routed in $C$. Assume the procedure Assign_wire of Fig. 5.9 is used to assign wires of $W$ to rows of $C$. Let $n_i$ be the number of wires of $W$ beginning in column $i$, and let $o_i$ and $u_i$ be the number of rows in column $i$ that are occupied and unoccupied, respectively, before Assign_wire scans column $i$. Assume that for column $i$, the above procedure fails to assign each of the $n_i$ wires to some unoccupied row of $C$. Then it must be true that $u_i < n_i$. By definition, the number of occupied rows is given by $o_i = r - u_i$. Now our original assumption is that for all $i$, $r \ge m = o_i + n_i$. Eliminating $r$ in the last two relations and combining them yields $u_i \ge n_i$. This contradicts the implication that $u_i < n_i$. Therefore, the assertion that $r$ is the maximum number of wires in any column, and the assertion that the procedure fails to assign all wires in column $i$ to the $r$ rows, are contradictory; one or the other must be false. Hence, sufficiency is proved. $\square$

---

```
for col:= 1 to c do
    for row:= 1 to r do
        if row in col is unoccupied then
            begin
                Assign to row some unassigned wire i where l_i = col;
                Mark row as occupied between columns l_i and r_i.
            end;
```

---

**Fig. 5.9.** The procedure Assign_wire for assigning wires of $W$ to rows of $C$.

**Lemma 5.3:** The algorithm *HR-TrailTrace* computes the set HV containing one layout of each nondominating height vector $(U, L)$ from among all layouts of minimum width $W$ for any reordering $M'$ of $M$.

**Proof:** The proof is by induction. First, we show that the algorithm computes a minimum-width layout. This follows directly from Lemma 5.1, since PMCTC($M'$) contains every minimum-sized cover of each of the cover types of $M'$. We choose from PMCTC($M'$) a cover with a minimum number of trails with respect to all covers in PMCTC($M'$). Any permutation and orientation of the trails of this cover results in a minimum-width layout of $M'$.

Now we prove that *HR-TrailTrace* generates the set HV containing one layout of $M'$ for each nondominating height vector $(U, L)$ of minimal width $W$. The algorithm selects each minimum-sized cover $tc$ in the PMCTC($M'$) and generates systematically all layouts for each $tc$, all such layouts being of minimal width. The trails of $tc$ are permuted, and oriented. The closed dual trails are also broken exhaustively, and the height vector $(U, L)$ of each resulting layout is measured. The break point is selected by scanning the layout from left to right and finding the columns in the U- and L-regions, respectively, in which the number of intracell wire crossings is a maximum. The U- and L-regions can be routed in $U$ and $L$ rows, respectively, as implied by Lemma 5.2. A layout of $M'$ of minimum width having a height vector $(U_i, L_i)$ is placed in HV if there is no other layout of $M'$ of minimal width with height vector $(U_j, L_j)$ such that $(U_j, L_j)$ dominates $(U_i, L_i)$, and if there is no layout already in HV with a height vector $(U_k, L_k)$ such that $(U_k, L_k) = (U_i, L_i)$. There exists no minimum-width layout of $M'$ outside HV whose height vector dominates the height vector of any layout in HV. $\square$

**Theorem 5.1:** The *HR-TrailTrace* algorithm solves the WHR layout minimization problem exactly for any dual series-parallel circuit.

**Proof:** The theorem is proved by contradiction. Lemma 5.3 states that *HR-TrailTrace* computes the set HV of minimum-width layouts of the nondominating height vectors. A cell's height is given by $H = U + P + C + L + N$. Assume there exists a layout $Q$ of $M'$ of minimum width not in HV. Furthermore, let the height $H_q$ of layout $Q$ be less than the height $H_r$ for any layout $R$ in HV. Therefore, since $P, C$ and $N$ are the same for both $Q$ and $R$, we have $U_q + L_q < U_r + L_r$. This implies that either $U_q < U_r$ or $L_q < L_r$, or both. But if any of these is true, then $(U_q, L_q)$ dominates $(U_r, L_r)$, and therefore $Q$ is in HV. $\square$

## 5.3.4 Time Complexity

We now discuss the analytical worst-case time complexity of *HR-TrailTrace*, which is about $O(h^3 h^2)$, where $h$ is the height of $M$. As stated previously, the WHR layout problem is at least NP-hard [Ch91]. We derive the time complexity as follows from the algorithm in Fig. 5.5. In the main body of *HR-TrailTrace*, the input of $M$ requires $O(E)$ time, where $E$ is the number of edges in $M$. The construction of the composition tree $T$ from $M$ takes $O(E)$ time. The generation of abutment numbers has $O(E^2)$ complexity. Node_CTC($T$) requires $O(E|CTC|^2)$ time. Pruning Graph_CTC down to Min_covers takes $O(|CTC|)$ time, and the procedure Generate_min_height_layout has complexity $O(|Min\_covers|4^{E/4}E^{E+2}2^E)$.

The complexity of Generate_min_height_layout and Node_CTC dominates all the other procedures of *HR-TrailTrace*. First, we shall analyze Node_CTC. The **if-else** statement in the inner loop of the function Ordered_children_CTC compares a new cover to each cover in its equivalence class included in the Temp_children_CTC list; the size of an equivalence class is $O(CTC)$. Each comparison of covers requires $O(E)$ time. Therefore, the **if-else** statement has time complexity $O(E|CTC|)$. This statement is executed each time a new cover is created as a result of concatenating two smaller covers of subgraphs being merged; this occurs on the order of $|CTC|$ times. Therefore, the complexity of Node_CTC is $O(E|CTC|^2)$. The worst-case size of CTC is $O(h!^h 2^E)$, assuming that every concatenation of covers produces a unique cover. In a composition tree $T$ of a graph $M$ of height $h$, the degree of any node of $T$ cannot exceed $h$. The size of the CTC can grow by a factor no greater than $h!$ at any given node of $T$ through reordering. This is an upper bound on the number of times that Ordered_children_CTC can be invoked from each call of Node_CTC. Since $T$ has nodes of degree proportional to $h$, the number of such nodes of $T$ is $O(h)$. The number of new covers generated at these internal nodes of degree $h$ cannot exceed $O(h!2^h)$. The total number of covers generated by the $h$ nodes of $T$ is $O((h!2^h)^h) = O(h!^h 2^E) = O(2^{h^2} h^h)$.

The time complexity for Generate_min_height_layout is derived as follows. For a given circuit, the number of layouts generated by Generate_min_height_layout is given by the following formula:

$$|Layouts| \; = \; |Min\_covers| \times |breaks| \times |cover|! \times 2^{|cover|} \qquad (5.1)$$

For a given cover, $|breaks| = O(4^{E/4})$, $|cover|! = O(E^E)$, and $|reflections| = O(2^E)$. It takes $O(E^2)$ time to calculate the height of each layout. The maximum number of breaks of closed trails occurs when a cover of $E$ edges has a layout of $E/4$ closed dual trails, each of which can be broken independently, and has four edges. Therefore, the total number of possible breaks is $4^{E/4}$. The number of trail permutations is $|cover|!$, where $E$ is the upper bound on the size of a cover. Now $E! = O(E^E)$, and the number of different ways to orient or reflect $E$ trails is $2^E$.

We can simplify expression (5.1) as follows. We note that $E = O(h^2)$ and $|Min\_covers| = O(|CTC|) = O(h!^h 2^E)$. Expressing the worst-case time complexity solely in terms of $h$ yields $O(h!^h 2^{h^2} 4^{h^2/4} 2^{h^2} h^{2h^2+4}) \approx O(h^{3h^2})$. However, despite the rapid growth of this complexity function, *HR-TrailTrace* handles all circuits of practical size, as we show in the next section.

## 5.4  COMPLETE STUDY OF PRACTICAL CELLS

As in Chapter 3, we have performed a complete analysis of the 3503 practical-sized series-parallel circuits of height four or less. We implemented *HR-TrailTrace* in 7000 lines of Pascal code and ran the program on each of these circuits, collecting various data such as best- and worst-case cell height, cell width, area improvement due to height optimization, execution time, and the number of covers retained at various points in the program. The complete study took 220 hours on a MicroVax II. The results of this study are presented below.

**Height Distribution.** Figure 5.10 shows the distribution of the minimum layout heights for the entire class of practical circuits. The cell height $H$ is calculated according to the formula of $H = U + P + C + N + L$. The C-region always has a height of one in our layout methodology. The shortest layouts found have two metal rows in the U-region and two in the L-region, whereas the tallest ones have five and four rows in the two regions, respectively. By far the most common height is seven, with three rows each in the U- and L-regions.

Note that a height of five is the minimum possible for any circuit, so that all the layouts achieving this height are known to be optimal in total area. It is possible to determine a tight lower bound on cell height achievable regardless of width. Of course, if the lower bound for cell height is realized, and the cell is also of minimum width, then the cell has minimum area. Having examined about 30 single-cell layouts, we have not found any where increasing their width can reduce their height. Therefore, based on this small sampling, we conjecture that most or all of the layouts produced by *HR-TrailTrace* have not only the minimum height among all minimum-width cells, but have minimum height for cells of all widths.

| Minimum height $H$ | Number of circuits | Percent of circuits |
|:---:|:---:|:---:|
| 5 | 49 | 1.4 |
| 6 | 618 | 17.6 |
| 7 | 2400 | 68.5 |
| 8 | 342 | 9.8 |
| 9 | 90 | 2.6 |
| 10 | 4 | 0.1 |

**Fig. 5.10.** Layout height distribution of all practical series-parallel circuits.

**Area Improvement.** Figure 5.11 shows the area improvement distribution for the class of all practical circuits. In generating each layout of a circuit, a placement and orientation of the trails of each cover were chosen that yielded a layout of minimum height. The improvement parameter compares the tallest layout to the shortest layout among all layouts of minimum width for a given circuit, according to the following formula:

Area improvement =
  (Area of tallest layout - Area of shortest layout) / Area of shortest layout.

This is a measure of how much larger a minimum-width layout produced by *R-TrailTrace* might be than that of *HR-TrailTrace* in the worst case, since *R-TrailTrace* does not consider layout height at all. Layouts produced by heuristic methods such as the one in [Uv81] can be even further from optimal, since they generate layouts that are not optimal either in width or height. Note that very large area differences occur, with an average improvement being in the 20 to 30% range.

These data demonstrate the important role of height minimization in the overall area minimization problem. Whereas we show in Chapter 3 that width minimization can affect the cell area by as much as 20%, height minimization can reduce the area by more than 80%. Figure 5.12 shows two minimum-width layouts of a cell implementing the function $\neg(*(+a(*bc))(+de(*(+fg)(+hi))))$, where layout (a) is 83% larger than layout (b).

| Percent area improvement | Number of circuits | Percent of circuits |
| --- | --- | --- |
| 80-90 | 12 | 0.3 |
| 70-79 | 4 | 0.1 |
| 60-69 | 44 | 1.3 |
| 50-59 | 248 | 7.0 |
| 40-49 | 304 | 8.7 |
| 30-39 | 336 | 9.6 |
| 20-29 | 1484 | 42.4 |
| 10-19 | 796 | 22.7 |
| 0-9 | 275 | 7.8 |

**Fig. 5.11.** Best-case area improvement distribution among optimal-width layouts.

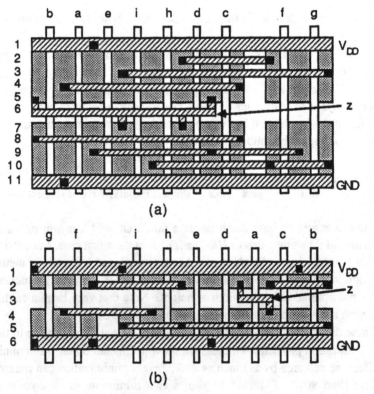

(a)

(b)

**Fig. 5.12.** Two minimum-width layouts of a cell implementing the function $\neg(*(+a(*bc))(+de(*(+fg)(+hi))))$: (a) layout of height 11; (b) layout of height 6.

| Paper | Figure | Function | Paper's height (U,L) | Our height (U,L) | Height impro- vement | Width impro- vement | Run time (s) |
|---|---|---|---|---|---|---|---|
| [UC81] | 4b,5 | ¬(+a(*bc)(*de)) | (3,2) | (3,2) | 0 | 0 | 6 |
| | 11a | ¬(*a(+b(*c(+d(*e(+f(*g(+hi)))))))) | (3,3) | (3,3) | 0 | 0 | 50 |
| | 14 | ¬(+(*ab)(*(+cd)(+ef))(*gh)) | ------ | (3,3) | --- | 12% | 12 |
| [KW85] | 12 | ¬(+(*ab)(*c(+de))) | (3,2) | (3,2) | 0 | 0 | 2 |
| [MH87] | 13 | ¬(+(*ab)(*(+cd)(+ef))(*gh)) | (4,3) | (3,3) | 14% | 0 | 12 |
| [MH90] | 14b | ¬(+(*(+(*ih)(*gf)e)(+dc)b)a) | (3,3) | (3,3) | 0 | 0 | 12 |
| [HS88] | 9 | ¬(+jk(*hi)g(*ef)(*d(+c(*ab)))) | (3,4) | (3,3) | 14% | 18% | 135 |
| [CH88] | 1 | ¬(+a(*bc)) | (3,3) | (2,2) | 40% | 0 | 1 |
| | 2 | ¬(+a(*bc)d) | (2,2) | (2,2) | 0 | 0 | 2 |
| | 9c | ¬(*(+ab)(+c(*de))(+if(*jk)(*gh))) | (4,4) | (3,3) | 29% | 0 | 120 |
| | 10 | ¬(+(*abc)(*(+de)(+fg))(*hi)) | (4,3) | (3,3) | 14% | 0 | 18 |
| | 10 | ¬(+(*(+fl)(+(*(+nx)y)(*nm)))(*ij(+hc))) | (4,3) | (4,3) | 0 | 0 | 58 |
| | 10 | ¬(+(*ah)(*(+bcd)(+kmn))) | (3,3) | (2,3) | 17% | 0 | 10 |
| | 10 | ¬(+(*c(+ab)(+ln))(*mx)(*hy)) | (3,3) | (3,3) | 0 | 0 | 24 |
| | 10 | ¬(+(*lm(+(*jk)xy))(*fg)) | (3,3) | (3,2) | 17% | 0 | 29 |
| | 10 | ¬(+(*(+abcde)(+fg))(*hi)) | (3,3) | (2,3) | 17% | 0 | 3 |
| [MD88] | 1 | ¬(+(*ab)c(*de)) | (3,2) | (3,2) | 0 | 0 | 6 |
| | Table1 | ¬(*a(+b(*c(+d(*e(+f(*g(+hi))))))))) | 6 | 6 | 0 | 0 | 50 |
| | Table1 | ¬(+(*ab)(*(+cd)(+ef))(*gh)) | 6 | 6 | 0 | 0 | 12 |
| | Table1 | ¬(*e(+(*ab)(*cd))) | 5 | (3,2) | 0 | 0 | 12 |
| [KK88] | 1b | ¬(*(+ab)c(+de)) | (2,3) | (2,3) | 0 | 0 | 6 |
| | 5a | ¬(+(*ab)(*(+cd)(+ef))(*gh)) | (4,3) | (3,3) | 14% | 0 | 12 |
| | 5b | ¬(+(*abcd)(*efgh)(*(+ij)(+kl)(+mn)(+op))) | (4,5) | (3,4) | 25% | 0 | 280 |
| [Ma89] | 3d | ¬(*g(+a(*(+bc)(+d(*ef))))) | (3,3) | (3,3) | 0 | 0 | 11 |
| [LC89] | 5 | ¬(+(*ab)(*(+cd)(+ef))(*gh)) | (3,3) | (3,3) | 0 | 0 | 12 |
| [Ma85] | 15b | ¬(+a(*(+bc)(+de))) | (2,3) | (2,3) | 0 | 0 | 23 |

**Fig. 5.13.** Comparison of width and height between optimal layouts by *HR-TrailTrace* and those of all single series-parallel functional cell layouts in published papers.

**Area Improvement Comparison.** Figure. 5.13 shows the results of applying *HR-TrailTrace* to a large number of published cell designs. We found that in about half the cases, it generated significantly smaller layouts than previous methods. Unlike the other methods cited, *HR-TrailTrace* always generates optimal layouts for all practical-sized circuits. Note that the examples of Fig. 5.13 were presumably chosen by their authors to show their methods at their best. Most of these circuits have circuit height four or less, but as Fig. 5.13 shows, *HR-TrailTrace* can also efficiently handle circuits from [CH88] that are of height five and six.

**CTC Size.** Figures 5.14 and 5.15 show how many trail covers are generated by the trail covering procedure Node_CTC and are passed to the layout procedure Generate_min_height_layout, respectively. Figure 5.14 shows, for all circuits of practical size, the distribution of the number of covers in Graph_CTC, the PMCTC in the *HR-TrailTrace* procedure in Fig. 5.5. Although our analytical bound on the number of covers of $M$ in the final CTC is $O(h^{2h^2})$, Fig. 5.14 shows that this set has fewer than 400 covers for almost all circuits examined, with none of the practical-sized circuits having more than 1464 covers.

The set of covers Graph_CTC is reduced to Min_covers, a set of all covers of $M$ that have the minimum size regardless of cover type. The distribution of the sizes of the set Min_covers is shown in Fig. 5.15. This set has fewer than 100 covers for the vast majority of circuits (89%), with 460 covers being the upper limit. *TrailTrace* and *R-TrailTrace* have a fixed upper limit of 42 on the number of covers required to ensure an optimal-width layout. *HR-TrailTrace* does not have such a fixed limit, and our worst-case analysis presented above suggests that the number of covers may be prohibitively large. However, our experimental data show that for circuits with $h \leq 4$ in the worst case, the number of covers in Graph_CTC of *HR-TrailTrace* is about 35 times larger than in Graph_CTC of *R-TrailTrace*, and that of Min_covers is about 10 times larger. This is a modest increase compared to the large potential area improvement offered.

**Layouts Placed and Routed.** The number of layouts placed and routed by *HR-TrailTrace* is related to the number of covers in Min_covers, the number of breaks of closed dual trails, and the number of trails in a cover. For circuits with $h \leq 4$, the maximum number of covers in Min_covers is 460 according to Fig. 5.15, and the maximum number of ways to break the closed trails is $4^4 = 256$. Furthermore, the data presented in Fig. 3.12(b) show that the maximum number of trails required to cover $M$ with height $h \leq 4$ is four. Therefore, the maximum number of trail permutations in the layout phase is $4! = 24$, and the maximum number of trail reflections is $2^4 = 16$. Based on these data, we can compute an upper bound on the number of layouts generated for any circuit of $h \leq 4$. As shown above, the number

| Number of trail covers in Graph_CTC | Number of circuits | Percent of circuits |
|---|---|---|
| 1-399 | 3193 | 91 |
| 400-999 | 298 | 8.5 |
| 1000-1464 | 12 | 0.5 |

**Fig. 5.14.** Distribution of Graph_CTC's size over all practical series-parallel functional cells.

| Number of trail covers in Min_covers | Number of circuits | Percent of circuits |
|---|---|---|
| 1-99 | 3129 | 89.3 |
| 100-399 | 368 | 10.5 |
| 400-460 | 6 | 0.2 |

**Fig. 5.15.** Distribution of Min_cover's size over all practical series-parallel functional cells.

of layouts placed and routed in any circuit is given by the following formula:

$$|Layouts| = |Min\_covers| \times |breaks| \times |cover|! \times 2^{|cover|}$$
$$= 460 \times 256 \times 24 \times 16$$
$$= 45,219,840$$

Figure 5.16 shows the distribution of actual number of layouts routed for the 3503 practical circuits. Most circuits require fewer than 1000 layouts to be evaluated. This number grows to a maximum of 811,008 layouts in the worst case. Typical cases are much better than this worst-case analysis indicates; see Fig. 5.16. For example, 95% of the set of 3503 circuits have optimal covers containing two or fewer trails, and almost all the rest have three trails. Only 0.2% of covers of circuits in this class have four trails. We make no attempt in *HR-TrailTrace* to use bounding and search reordering techniques to reduce these numbers, because the typical run times are small enough. However, we believe that these worst-case numbers can be dramatically reduced by applying such speedup techniques.

| Number of layouts generated | Number of circuits | Percent of circuits |
|---|---|---|
| 1-1000 | 2899 | 83 |
| 1000-10,000 | 460 | 13 |
| 10,000-100,000 | 132 | 4 |
| 100,000-1,000,000 | 12 | 0.3 |

**Fig. 5.16.** Distribution of layouts generated for all practical-sized series-parallel circuits.

| Execution time (min) | Number of circuits | Cumulative percent of circuits |
|---|---|---|
| 0-1 | 1345 | 38 |
| 1-2 | 770 | 60 |
| 2-3 | 346 | 70 |
| 3-4 | 270 | 78 |
| 4-5 | 156 | 82 |
| 5-6 | 106 | 85 |
| 6-7 | 62 | 87 |
| 7-8 | 74 | 89 |
| 8-9 | 60 | 91 |
| 9-10 | 46 | 92 |
| 10-20 | 206 | 98 |
| 20-30 | 38 | 99.30 |
| 30-40 | 10 | 99.60 |
| 40-50 | 2 | 99.70 |
| 62 | 2 | 99.75 |
| 65 | 2 | 99.80 |
| 71 | 2 | 99.85 |
| 94 | 2 | 99.90 |
| 208 | 2 | 99.95 |
| 504 | 2 | 100 |

**Fig. 5.17.** Execution time distribution of *HR-TrailTrace* for all practical series-parallel circuits.

**Actual Execution Time.** Despite the worst-case complexity calculated above, we expected that the actual complexity would be much less for practical circuits. We gathered the actual execution times of *HR-TrailTrace* on the whole class of practical-sized series-parallel circuits. It is important to note that relatively inexpensive workstations are available that offer about two orders of magnitude more performance than the MicroVAX II used in these experiments.

The data are presented in Fig. 5.17. About 90% of these circuits ran in 8 minutes or less, and 98% of them ran in less than 20 minutes; the mean execution time for all the circuits is about 3 minutes. A few circuits required more than 10 minutes. The slowest two circuits required about 8 hours each, four took from 1.5 to 3 hours each and all the rest completed execution in an hour or less. Almost all of this additional time is spent in generating layouts from covers. These experimental data confirm our hypothesis that such an algorithm can indeed be practical. Moreover, on the more powerful workstations now available, the average time should be about a second, and even the worst-case times should reduce to just minutes.

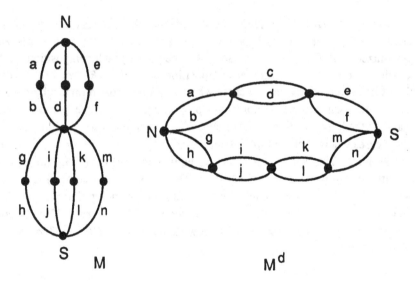

**Fig. 5.18.** The graph model of the circuit implementing the function $z = \neg(*(+(*ab)(*cd)(*ef))\,(+(*gh)(*hi)(*jk)))$.

Figure 5.18 shows the graph model of the circuit implementing the function $z = \neg(*(+(*ab)(*cd)(*ef))(+(*gh)(*hi)(*jk)))$, which is one of the two circuits whose layout is generated in about 3 hours by *HR-TrailTrace*; the other one is the dual of this circuit. This circuit has one of the worst-case execution times, requiring the generation of 368,640 layouts. It has 48 minimum-sized covers, each of which has four trails. Some of these covers have up to three closed dual circuits, each with four edges. Equation (5.1) gives the number of layouts generated in the worst-case cover for this circuit.

$$
\begin{aligned}
\text{|Layouts for cover|} \quad &= \quad \text{|breaks|} \times \text{|cover|!} \times 2^{\text{|cover|}} \\
&= \quad (4 \times 4 \times 4) \times 4! \times 2^4 \\
&= \quad 24{,}576
\end{aligned}
$$

Forty-eight such covers require a total of nearly one million layouts to be generated. Only a few circuits like this have the worst-case number of closed dual circuits and trails in a cover. The vast majority of practical-sized circuits have far fewer layouts generated for them, as Fig. 5.16 shows. Applying the bounding techniques outlined above would drastically reduce the execution time of these worst-case circuits.

## 5.5  PLANAR CELL LAYOUT

The *HR-TrailTrace* algorithm for dual series-parallel circuits can be generalized to handle dual planar circuits in a way quite analogous to the way *TrailTrace* is generalized to *P-TrailTrace* in Chapter 4. The planar graph composition method is substituted for the series-parallel composition method. *HR-TrailTrace* performs circuit graph reordering only in cases where multiple two-terminal subgraphs are in series. In such situations, nonisomorphic subgraphs can be permuted without changing the logic function of the circuit. The planar graph composition method can detect cases where such multiple two-terminal subgraphs are in series and reordering can be done. The method for detecting equivalent trail covers in the generalized algorithm is the same as in *P-TrailTrace*. The layout generation method is the same as used in *HR-TrailTrace*, namely the permuting and orienting of dual trails. Overall, the generalization is a blending of *P-TrailTrace* and *HR-TrailTrace*. The performance is expected to be comparable to, but somewhat slower than, *HR-TrailTrace* on series-parallel circuits.

# CHAPTER VI

# CELL ARRAY WIDTH AND HEIGHT
# MINIMIZATION

In this chapter, we extend our layout theory to arrays of two or more cells. We first define a practical cell-array layout style and present an algorithm *HRM-TrailTrace* that generates cell arrays optimal in width and height in our layout style. We prove this optimality claim and present experimental results to substantiate it [MH91].

## 6.1 LAYOUT PROBLEM

We define the cell-array layout style and a corresponding layout optimization problem in this section. A one-dimensional *cell array* is a linear array of two or more interconnected functional cells. The cell-array layout style is illustrated by the two-cell example in Fig. 6.1. This style is defined by the 16 assumptions of Fig. 5.2 plus two new ones, *assumptions 17* and *18*, stated as follows:

17.  Assign all intercell connections to metal wires in the U-region.
18.  Group all transistors of each cell together.

We discuss below these assumptions, along with other key assumptions of Fig. 5.2, in the context of cell-array layout.

According to assumption 7, diffusion is used to make vertical interconnections between source or drain transistor terminals and horizontal metal. Hence, in our cell-array style, vertical diffusion is used to connect the output of a cell to horizontal metal for intercell routing. This is illustrated in Fig. 6.1, where vertical diffusion is used between columns *d* and *e* in row 3 to route the output of cell 1 (signal *i*) from row 4 to row 3. As in Chapter 5, we emphasize that vertical diffusion could be replaced with metal in a two-metal routing scheme. Intercell routing is done solely in the U-region according to assumption 17, and is shown in rows 2 and 3 in Fig.

6.1. Diffusion abutment of neighboring functional cells via the $V_{DD}$ and GND terminals is used whenever possible, thus reducing the width of the combined cell by one column width. The cells of Fig. 6.1 abut in this way between columns $f$ and $g$ in rows 4, 5 and 7. The 12 transistors of cell 1 in Fig. 6.1 are grouped together to the left of the eight transistors forming cell 2 in the same figure, according to assumption 18.

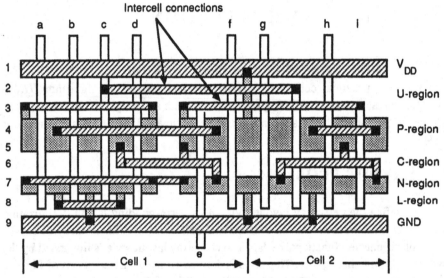

**Fig. 6.1.** An illustration of our cell-array layout style.

We now discuss the justification for assumption 17 in the context of other functional cell layout methods. Although no other method limits intercell routing solely to the U-region, many use this region extensively, for example the standard cell styles of [Uv81, Ma85, MD88, Ma89, CH88]. Others use the C-region solely for vertical intracell connections, as we do, relegating all horizontal connections to the U- and L-regions [HH90, BA88]. TOPOLOGIZER uses both the U- and C-regions for intercell routing [KW85]. Sc2 confines all intracell and intercell routing to the C-region [Hi85]; this is a precedent for using a single routing region in a cell-array layout style. Therefore, this assumption is in keeping with the spirit of many other layout systems.

Based on the foregoing style constraints, we now define the corresponding cell-array layout optimization problem formally. The cell-array height $H$ is given by the following equation

$$H = \max\{U_i\} + P + C + N + \max\{L_j\}$$

where $P$, $C$ and $N$ are the number of rows in the P-, C- and N-regions of the entire cell array. These numbers are the same for all cells, according to assumption 10 of Fig. 5.2; this is shown in Fig. 6.1. Max$\{U_i\}$ and max$\{L_j\}$ are the numbers of rows in cells $i$ and $j$ with the largest U- and L-regions, respectively, as shown in Fig. 6.1. The cell-array width $W$ is given by the equation

$$W = T + G + 1$$

where $T$ is the number of transistors in either the pullup or pulldown subcircuit of all the functional cells in the array, and $G$ is the number of diffusion gaps present. We define the *WHRM layout minimization problem* as follows: given a set of dual series-parallel circuits with circuit graphs $\{M_1,..., M_n\}$ and placement $P$ of the $n$ cells in the cell array, find a layout in the style of Fig. 6.1 with minimum $H$ from among all layouts of minimum $W$, for any set of reorderings $\{M'_1,..., M'_n\}$ of $\{M_1,..., M_n\}$.

## 6.2 HRM-TRAILTRACE ALGORITHM

In this section we present the algorithm *HRM-TrailTrace*, a generalization of *HR-TrailTrace* for multiple cells, that solves the WHRM optimization problem exactly. First, a brief description of the algorithm is given, followed by a design example. We then prove the optimality of *HRM-TrailTrace*. Finally, we analyze the algorithm's time complexity.

### 6.2.1 Description

The algorithm that computes the optimal one-dimensional cell array as described above for a given placement of functional cells is given in Fig. 6.2.

A brief description of *HRM-TrailTrace* follows. First, the composition expression for each cell is read. Next, the input and output signals that cross the left and right boundaries of each cell are assigned to the proper cell boundary, thus decomposing the cell-array layout problem into several single-cell layouts with boundary constraints. The procedure Generate_min_height_layouts2(cell,H) generates each nondominated optimal cover and layout for a given cell. This procedure is called for each cell in the array. It is essentially the same as the *HR-TrailTrace* procedure in Fig. 5.5, with the following difference. Generate_cell_layouts in *HRM-TrailTrace* calls the procedure Generate_min_height_layouts2 instead of Generate_min_height_-layouts, which *HR-TrailTrace* uses. Generate_min_height_layouts2 is Generate_-min_height_layouts shown in Fig. 5.5 with the addition of statement 1. This new procedure permutes the isomorphic subgraphs affecting layout height in the cell-array style of Fig. 6.1. (A complete description of Generate_min_height_layouts is given in Chapter 5.)

---

**procedure** Find_optimal_cell_array(HV,Smallest_H);
**begin**
for each j do {initialize MHV} MHV:= $HV_{1,j}$;
  for cell:= 2 to n do
    **begin**
    for each j do
        for each k do {Calculate next vector}
            **begin**
            HVtemp := lub{$MHV_k$, $HV_{cell,j}$};
            Save HVtemp in MHVtemp if not dominated or equalled by a vector in MHVtemp;
            **end**;
        for each vector in MHVtemp do {Update MHV} MHV:= MHVtemp;
        **end**;
Smallest_H:= $\infty$;
for each k do {Find smallest overall layout}
    if $MHV_k$ < Smallest_H then Smallest_H:= $MHV_k$;
**end**; { Find_optimal_cell_array}


**procedure** Generate_cell_layouts(Cell,HV); {This is essentially HR-TrailTrace}
**begin**
Construct composition tree T from cell's M;
Identify_isomorphic_subgraphs;
Graph_CTC:= Node_CTC(T);
if any minimum size cover has a $V_{DD}$/GND terminal pair **then**
    Min_covers:= Retain for layout all trail covers of min. size from Graph _CTC;
**else** Min_covers:= Retain for layout all trail covers of min. size and min. size+1
               from Graph _CTC;
Generate_min_height_layout2(Cell, Min_covers, HV);
**end**; { Generate_cell_layouts}


**procedure** HRM-TrailTrace;
**begin**
Input_cells;
Calculate_boundary_signals;
for each Cell do Generate_cell_layouts(Cell,HV)
Find_optimal_cell_array(HV,Smallest_H);
**end**; { HRM-TrailTrace }

---

**Fig. 6.2.** The *HRM-TrailTrace* algorithm.

Finally, the height vectors $(U_i, L_i)$ produced for each cell by Generate_cell_-layouts are processed by Find_optimal_cell_array, which computes the optimal cell-array layouts from the height vectors of the individual cells. In Find_optimal_cell_array, the following least upper bound (lub) operation is performed on two adjacent layouts to determine the minimum height of the layout resulting from combining them. Assume the height pair of one layout is $(A,B)$ and that of the other is $(C,D)$. Then their least upper bound is

$$\text{lub}\{(A, B), (C, D)\} = (\max\{A, C\}, \max\{B, D\})$$

Let $H_{ij}$ be the jth height-pair of cell $i$. Let $H_j$ be the jth height-pair of a undominated layout composed of the $i$-1 cells to the left of cell $i$. The overall layout with least height (smallest_H) for the $n$ functional cells is calculated by procedure Find_optimal_cell_array, as shown in Fig. 6.2.

## 6.2.2.  Design  Example

We now illustrate the *HRM-TrailTrace* algorithm by an example. Figure 6.3 shows a circuit composed of three series-parallel subcircuits $C_1, C_2$ and $C_3$ and their graph models $M_1, M_2$ and $M_3$, respectively. The algorithm's input data consists of these three circuits, each of which will be implemented as a cell in a 3-cell array. The procedure Calculate_boundary_signals of Fig. 6.2 assigns signals shared by two or more cells to the appropriate side(s) of the cells. For example, cell 1 implementing $C_1$ shares signal $b$ with cell 2 implementing $C_2$; therefore, signal $b$ is assigned to the right side of cell 1 and the left side of cell 2, as shown in Fig. 6.4.

The procedure Generate_cell_layouts is applied to each cell successively. This procedure is much like *HR-TrailTrace* described in Chapter 5. The composition tree $T$ is built for the cell, and is used to identify all isomorphic subgraphs of the cell used later in the layout procedure. In the case of cell 2, for example, edges $b$ and $c$ are its isomorphic subgraphs. The PMCTC called Graph_CTC in Fig. 6.2 is calculated for the cell by procedure Node_CTC. Graph_CTC is transformed into the set Min_covers by pruning it of all covers of greater than minimum size, with the following exception. If no cover of minimum size is of a type with an endpoint pair $(S, N)$, as in the type $(S, N)/(S, S)$, which represents $V_{DD}$ and GND in the circuit, then all covers that are of minimum size plus one with an $(S, N)$ endpoint pair are also included in Min_covers. This exception ensures that the minimum height cell array for all arrays of minimum width is found, since these larger covers also can lead

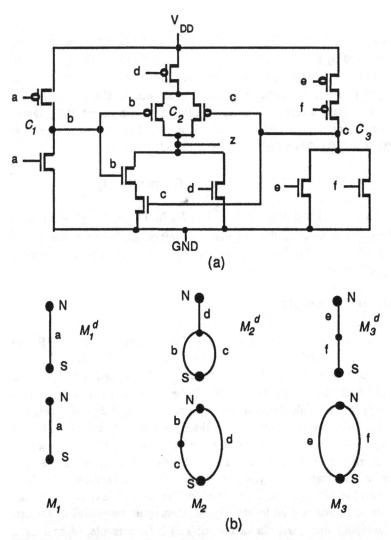

(a)

(b)

**Fig. 6.3.** An implementation of the function $z = \neg(+ d(*(\neg a)\neg(+ef)))$: (a) circuit and (b) graph model.

to minimum-width cell-array layouts by eliminating a diffusion gap between neighboring cells. Figure 6.5(a) shows two cells: cell 1, a layout with $V_{DD}$ and GND contacts on its right side corresponding to the (S,N) terminals of its cover, and cell 2, a minimum-width layout of a circuit cover without the (S,N) terminals. Cells 1 and 2 of Fig. 6.5 form a cell array of Fig. 6.5(b) with the resulting gap between

| V DD                   V DD | V DD                      V DD | V DD                       V DD |
|                         b | b                          c | c |
|         Cell  1 |         Cell  2 |         Cell  3 |
| GND              GND | GND                GND | GND                 GND |

**Fig. 6.4.** The 3-cell array illustrating the intercell boundary signals.

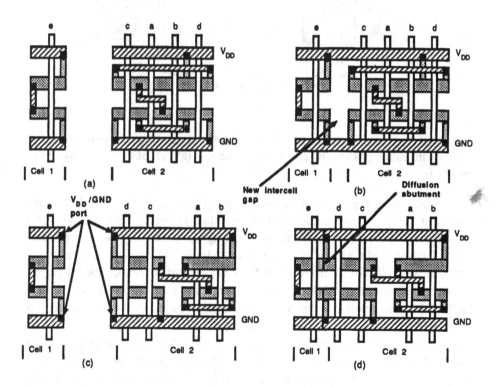

**Fig. 6.5.** The abutment of two cells forming an array: (a-b) intercell gap required; (c-d) diffusion abutment needing no gap.

them shown in the figure. Cell 2 in Fig. 6.5(c) is a layout of the same circuit as cell 2 in Fig. 6.5(a), but its cover has the (S,N) terminals defined by the $V_{DD}$ and GND contacts on its left boundary. Note that cell 2 of Fig. 6.5(c) is wider by one column

than the cell 2 in Fig. 6.5(a). However, the terminals of cell 1 abut the terminals of
cell 2, forming the 2-cell array of Fig. 6.5(d), with the same width as the array in
Fig. 6.5(b).

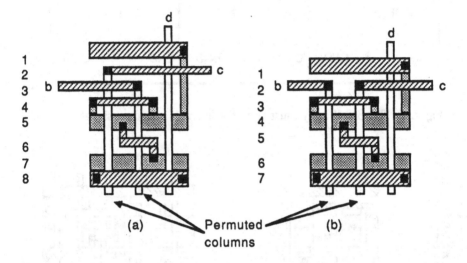

**Fig. 6.6.** Two layouts of circuit $C_2$ of Fig. 6.3(a) showing the permutation of
isomorphic subgraphs and its affect on cell height: (a) cell of height 8; (b) cell of
height 7.

At this point, the procedure Generate_min_height_layouts2 is applied to the
covers of the circuit; an outline of this procedure appears in Fig. 5.5. Here we
emphasize its differences with Generate_min_height_layouts. Layouts are generated
for each cover of Min_covers by breaking closed trails, and permuting and reflecting
the trails, as in *HR-TrailTrace*, with the following exception. Isomorphic subgraphs,
which are not permuted when covering the reorderings of the circuit in procedure
Node_CTC, are permuted at this point. Figure 6.6(a) illustrates this permutation for
edges $b$ and $c$ of $C_2$. It shows that the layout of cell 2 has a height vector $(U,L) =$
$(4,2)$ if $b$ and $c$ are not permuted, but the layout of Fig. 6.6(b) has the smaller height
vector $(3,2)$ when $b$ and $c$ are swapped. *HRM-TrailTrace* permutes all isomorphic
subgraphs in the layout generation procedure in an efficient manner, recognizing
cases where such permutations cannot affect cell height. For example, if the edges $b$
and $c$ of Fig. 6.6(b) were connected to the same side of the cell 2, instead of different
sides, then their permutation could not affect cell height; therefore, these edges would
not be permuted. This analysis keeps the number of such permutations small. The
set $HV_i$ of nondominated height vectors is calculated over all the layouts of a cell $i$.

| Height vector | $HV_1$ | $HV_2$ | $HV_3$ |
|:---:|:---:|:---:|:---:|
| 1. | (3, 2) | (4, 2) | (3, 2) |
| 2. | (2, 3) | (3, 4) | (2, 3) |

**Fig. 6.7.** A set HV of height vectors for the three cells of Fig. 6.4.

After the procedure Generate_cell_layouts has calculated HV for each cell, Find_optimal_cell_array selects layouts from all the cells that together form the optimal layout for the cell array. Assume HV containing the nondominating height vectors for the three cells of Fig. 6.4 is as shown in Fig. 6.7. An optimal selection of layouts represented by these height vectors in Fig. 6.7 for the 3-cell array is calculated as follows. The set of optimal height vectors for the subarray of cells 1 through $n$ is denoted $MHV_n$. For the case of cell 1, $MHV_1 = HV_1$. For $MHV_n$, the lub is calculated, pairing each vector of $MHV_{n-1}$ with each vector of $HV_n$. For example, for $n = 2$,

$$(4, 2) = \text{lub} \{(3, 2), (4, 2)\}$$
$$(3, 4) = \text{lub} \{(3, 2), (3, 4)\}$$
$$(4, 3) = \text{lub} \{(2, 3), (4, 2)\}$$
$$(3, 4) = \text{lub} \{(2, 3), (3, 4)\}$$

Since (4, 2) dominates (4, 3), $MHV_2$ is $\{(4, 2), (3,4)\}$, where neither vector dominates the other. For $MHV_3$, the calculation is as follows

$$(4, 2) = \text{lub} \{(4, 2), (3, 2)\}$$
$$(4, 3) = \text{lub} \{(4, 2), (2, 3)\}$$
$$(3, 4) = \text{lub} \{(3, 4), (3, 2)\}$$
$$(3, 4) = \text{lub} \{(3, 4), (2, 3)\}$$

Therefore, $MHV_3$ is $\{(4, 2), (3, 4)\}$. The height vector (4, 2) represents the smaller of the two cell-array layouts. This minimum-height cell array consists of the layouts corresponding to height vector 1 of each cell of Fig. 6.7. These three cell layouts result in minimum height for the cell array, namely $4 + 1 + 2 = 7$, since the C-region is always of height one. The final optimal cell-array layout is shown in Fig. 6.8, where cells 2 and 3 are connected by diffusion abutment.

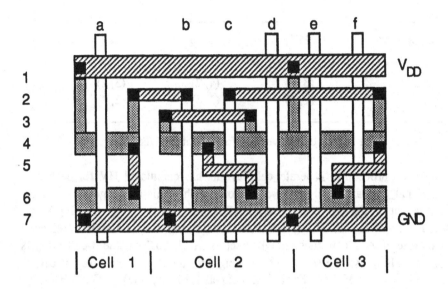

**Fig. 6.8.** Cell-array layout of the circuit of Fig. 6.3(a).


## 6.2.3  Optimality

We prove the optimality of the *HRM-TrailTrace* algorithm in this section. Three lemmas are developed as the basis for proving Theorem 6.1, which asserts the main optimality claim.

**Lemma 6.1:** The algorithm *HRM-TrailTrace* computes the set HV containing one layout of each nondominating height vector $(U, L)$ from among all layouts of minimum width $W$, for any reordering $M'$ of each cell $M$ of the cell array.

**Proof:** The proof is by induction, and the argument is essentially the same as that of Lemma 5.3, with one minor change, which we now discuss. The procedure GenerateMinHeightLayout2 of *HRM-TrailTrace* differs from GenerateMinHeight-Layout of *HR-TrailTrace* only in the inclusion of statement 1 in Fig. 5.5. This procedure selects each minimum-sized cover $tc$ in PMCTC($M'$) and generates systematically all layouts for each $tc$, each such layout being of minimum width. The trails of $tc$ are permuted and oriented, the dual closed trails are broken in each place possible, and the trail edges of the isomorphic subgraphs of $M'$ of $tc$ are permuted systematically. This permutation of isomorphic subgraph edges ensures that each layout, whose height can be affected by the permutation of its metal rows in the U-region that are connected to polysilicon columns, is generated and its height is measured. Figure 6.6 illustrates cell height reduction by the permutation of subgraphs $b$ and $c$ of the graph model $M_2$ in Fig. 6.3(b) of circuit $C_2$ from Fig. 6.3(a). □

**Lemma 6.2:** The procedure Find_optimal_cell_array(HV, smallest_H) in Fig. 6.2 computes a layout of minimum height $H$ for the cell array, given a set of nondominated height vectors for each cell of a cell array, and a placement of the cells.

**Proof:** A height vector $(U, L)$ of a layout of a cell contains the heights of each of its independent routing regions. The procedure Find_optimal_cell_array selects the layouts of the cells that result in the cell array of minimum height. This procedure begins with cell 1. At each step, it applies the lub operation pairwise to the nondominating vectors of the cell subarray composed of cell 1 through cell $i$-1 and the nondominating vectors of cell $i$. The lemma is proved by induction. For the one-cell case, the procedure selects the minimum-height vector from the set HV of height vectors for that cell. This vector is surely an element of $HV_1$ according to Theorem 5.1. Assume that Find_optimal_cell_array correctly computes the case of $n$-1 cells. We now prove that it does so correctly for $n$ cells. The procedure applies the least upper bound operation pairwise, pairing each vector in $MHV_{n-1}$, the set of nondominated height vectors for a cell array of the leftmost $n$-1 cells, with each vector of cell $n$ in $HV_n$.

The least upper bound is used by Find_optimal_cell_array because it corresponds to expanding the smaller of the corresponding routing regions to align it with the larger, so that their tops and bottoms coincide. The resulting vector corresponds to the cell array that has routing regions, each of which is tall enough to accommodate all the rows of horizontal wires of any of its constituent cells, because the lub operation applied to the vectors of neighboring cells is an *upper* bound. Figure 6.1 illustrates this alignment. For instance, the L-region of cell 1 has the two rows 8 and 9, whereas the L-region of cell 2 only needs to be one row high, since row 8 is empty. But when these cells are placed side-by-side in a cell array, the L-region of cell 2 is expanded so that the tops and bottoms of the L-regions of both cells coincide. This proves that the routing regions defined by the height vector are large enough for the cell interconnections. It is also true that each of the routing regions is no larger than is necessary to accommodate all the routing rows of each of the cells in the array. This follows from the fact that the lub operation that determines the size of the routing regions of the cell array is a *least* upper bound. Therefore, the cell array is as tall as, but no taller than, required by the nondominating vectors of each of the cells in the cell array.

Let the $k$th height vector of cell $n$ be $(U_{n,k}, L_{n,k})$. The vectors of $MHV_{n-1}$ represent the nondominated vectors of the cell subarray composed of the leftmost $n-1$ cells. Find_optimal_cell_array applies the least upper bound operation to every vector of $MHV_{n-1}$ paired with every vector of cell $n$ as follows.

$$\text{lub}\{(U_{MHV_i}, L_{MHV_i}), (U_{n,k}, L_{n,k})\} = (\max(U_{MHV_i}, U_{n,k}), \max(L_{MHV_i}, L_{n,k}))$$

Each resulting vector is placed in MHV_temp if no vector in MHV_temp dominates or equals it. Upon completion of the procedure, MHV_temp contains all nondominated height vectors for the array of $n$ cells. This proves that Find_optimal_cell_array computes all nondominating layouts of the cell array. The height vector of minimal height can be selected from MHV_temp according to Theorem 5.1. This inductive proof applies to a cell array of any size. Therefore, the minimum-height argument applies to the entire array. □

**Lemma 6.3:** Given a set of nondominated height vectors for each cell of a cell array, and a placement for the cells, the *HRM-TrailTrace* algorithm selects a layout for each cell that yields a cell-array layout of minimum height for all layouts of minimum width.

**Proof:** The proof is again by induction. Assume that the algorithm calculates a PMCTC($M'$) for each cell $M$ according to Lemma 5.1. The algorithm works correctly for a cell array consisting of one cell according to Theorem 5.1.

Consider an array of several cells. Assume that the algorithm has combined the leftmost $m$ cells such that the MHV containing the nondominated height vectors for all minimal width layouts for those $m$ cells has been found. We now show that the algorithm combines the layouts for cell $m+1$ such that the MHV contains the nondominated height vectors for all layouts of minimal width $W$ or of width $W + 1$ for those $m+1$ cells.

A cell layout may have a pair of contacts between a transistor source diffusion region and the $V_{DD}$ and GND lines along the cell side. The left side of cell 1 in Fig. 6.8 has contacts connecting metal to diffusion in rows 1 and 7. Such a pair of contacts along the side of a cell layout is called a *$V_{DD}/GND$ port*, as shown in Fig. 6.9(a). A port can be a *left* or *right* port depending on which side of the cell the corresponding pair of contacts appears. This pair of contacts is not to be confused with the $V_{DD}$ and GND boundary signals, shown in Fig. 6.4, that every cell has on both its right and left sides.

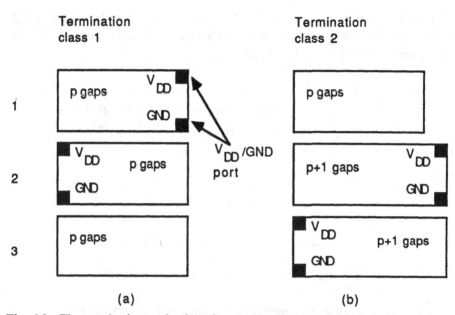

**Fig. 6.9.** The two basic termination classes: (a) class 1 and (b) class 2.

Two *termination classes* are based on the width of a circuit's narrowest layout with a $V_{DD}$/GND port. A circuit is of *termination class 1* if at least one of its possible layouts with a $V_{DD}$/GND port is of minimum width $w_{min}$. An HV of a circuit of termination class 1 has vectors of layouts only of width $w_{min}$. Termination class 1 is illustrated in Fig. 6.9(a), where all layouts in the HV are the same width, or alternatively, the layouts have the same number of diffusion gaps, since, for a given circuit, its layout width is directly related to the number of diffusion gaps. A circuit is of *termination class 2* if none of its layouts with a $V_{DD}$/GND port is of minimum width $w_{min}$. A circuit of termination class 2 has an HV in which no layout with a $V_{DD}$/GND port has minimum width. This circuit's HV contains height vectors corresponding to layouts of two widths, the minimal width $w_{min}$, and the near-minimum width $w_{min} + 1$. These cases correspond to layouts with $p$ gaps or $p + 1$ gaps, respectively. Layouts of both widths for a circuit of termination class 2 must be examined in constructing a cell array to ensure that the array has minimum height for *all* those of minimum width; see Fig. 6.5. This is because some minimum-width cell arrays can result from cells of width $w_{min} + 1$ due to diffusion abutment occurring between layouts of neighboring cells, as illustrated in Fig. 6.5. The 2-cell array in Fig. 6.5(b) is formed by cell 1 and 2 of Fig. 6.5(a), both minimum-width cells, whereas the array of Fig. 6.5(d), which is the same width as the one in Fig. 6.5(b), is formed from cell 1 and cell 2 of Fig. 6.5(c);

the latter cell is of width $w_{min}$ + 1. Termination class 2 is illustrated by Fig. 6.9(b), where layouts are included in the HV having $p$ gaps and no $V_{DD}$/GND port, and layouts with $p$ + 1 gaps and a $V_{DD}$/GND port .

| Case | Termination class of cell array | Termination class of right cell |
|:----:|:-------------------------------:|:-------------------------------:|
| 1. | 1 | 1 |
| 2. | 1 | 2 |
| 3. | 2 | 1 |
| 4. | 2 | 2 |

Fig. 6.10. The four cases of combining cells in Lemma 6.3.

Since the cell subarray formed by the leftmost $m$ cells and cell $m$+1 can be of either termination class, we must consider each of the four cases listed in Fig. 6.10. Figure 6.11 shows in detail how a cell combines with a cell on its right, illustrating the four cases of Fig. 6.10. The left and right cells have a minimum of $p$ and $q$ gaps, respectively. Figure 6.11 gives the cell width in terms of gaps and port type of the two cells, along with the resulting width and port of the combined cell shown in the righthand column of the same figure. The proof of the lemma is by exhaustive enumeration of each relevant combination of cells.

Case 1: The right and left cells are both of termination class 1. Figure 6.11(a) shows the two subcases resulting in a cell array of $p+q$ or $p+q+1$ gaps. We describe the ($p + q$)-gap subcase. The left cell of class 1 having $p$ gaps and a right $V_{DD}$/GND port is combined with a right cell of class 1, whose $V_{DD}$/GND port is on its left side. The resulting 2-cell array has $p+q$ gaps, with no new gaps created by this combination because the $V_{DD}$/GND ports of the two cells align to allow the cells to connect by diffusion abutment. In cases where the alignment of ports does not allow abutment, a new diffusion gap is created, increasing the width by one gap of the resulting array, as in the second subcase in Fig. 6.11.

Case 2: The right cell is of termination class 1 and left cell is of class 2. The three subcases resulting in cell arrays with $p+q+1$ or $p+q+2$ gaps are illustrated in Figure 6.11(b).

Case 3: The right cell is of termination class 2 and left cell is of class 1. The three possible combinations of these classes is shown in Fig. 6.11(c), and result only in

$p+q+1$ gaps. Note that only the resulting arrays of this case are of termination class 1, since some of the minimum-width layouts have a $V_{DD}$/GND port.

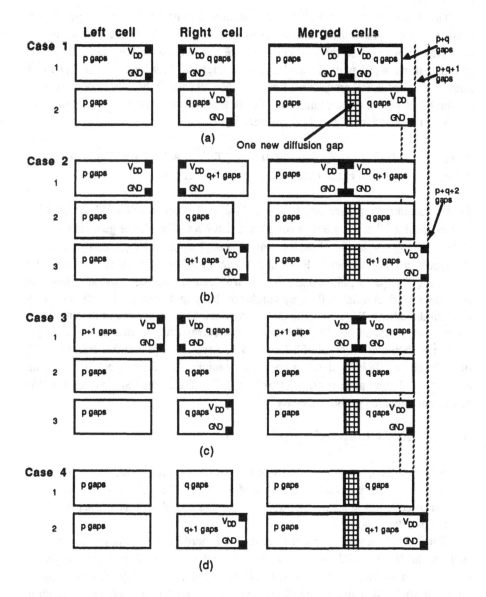

Fig. 6.11. An illustration of how two cells combine to form minimum-width cell arrays.

**Case 4:** Both the right and left cells are of termination class 2. Figure 6.11(d) shows the two subcases resulting in arrays having $p+q+1$ and $p+q+2$ gaps, respectively.

The above four case include all possible layouts of minimum width and minimum width plus one for the cell array of $m+1$ cells. The height vectors of the cell array of $m$ cells and those of cell $m+1$ are combined according to Lemma 6.2, resulting in an MHV with all nondominated height vectors for each port type and array width for the cell array of $m+1$ cells. When all cells in the cell array have been combined, a layout of minimum height is selected from all those of minimum width, as discussed in the proof of Theorem 5.1. $\square$

**Theorem 6.1:** The algorithm *HRM-TrailTrace* solves the WHRM layout minimization problem exactly for dual series-parallel circuits.

**Proof:** Lemma 5.1 proves that function Node_CTC computes the PMCTC($M'$) for any reordering $M'$ of $M$. Lemma 6.1 proves that *HRM-TrailTrace* generates the set HV containing one layout of each nondominating height vector $(U, L)$ from among all layouts of minimum width $W$ for any reordering $M'$ of each cell $M$, including the permuting of isomorphic subgraphs of $M$. Lemma 6.2 demonstrates that the procedure Find_optimal_cell_array combines the height vectors of each cell with those of the other cells to compute the nondominated height vectors of the entire cell array. The minimum-height cell-array layout corresponds to one of these height vectors as shown in the proof of Theorem 5.1. Lemma 6.3 proves that from the $HV_i$ of each cell $M_i$, *HRM-TrailTrace* computes a minimum-height layout from among all layouts of minimum width. Therefore, *HRM-TrailTrace* solves the WHRM layout minimization problem exactly. $\square$

## 6.2.4 Time Complexity

In this section, we analyze the time complexity of *HRM-TrailTrace*. First, we analyze the complexity as a function both of the height of the circuit graph and of the number of cells in the array. We also analyze the complexity as a function of the number of cells alone.

The WHRM layout problem is at least NP-hard [Ch91]. Figure 6.12 briefly outlines the *HRM-TrailTrace* algorithm along with the worst-case time complexity of its main procedures. It takes time $O(nE)$ to enter the $n$ circuit graphs representing an $n$-cell array. It requires time $O(nE)$ to decompose the problem by calculating which signals cross the left and right cell boundaries. The for loop in the main procedure of *HRM-TrailTrace* executes Generate_cell_layouts once for each cell. This procedure is essentially the *HR-TrailTrace* procedure analyzed in Chapter 5, with the modification noted above. The time complexity of this **for** loop is $O(n|\text{Min\_covers}|4^{E/4}2^{E}E^{E+2}(h!)^{h+1}) \approx O(nh^{4h^2})$. Finally, the height vectors of all the cells are combined to find the optimal layout of the cell array. The time complexity of this procedure is $O(nE^2)$. The time complexity of the entire *HRM-TrailTrace* algorithm is the complexity of the above-mentioned **for** loop, which is $O(nh^{4h^2})$. An analysis of the Generate_cell_layouts procedure is found in Chapter 5.

| HRM-TrailTrace | Complexity |
|---|---|
| **procedure** HRM-TrailTrace; | |
| Input_cells; | $O(nE)$ |
| Calculate_boundary_signals; | $O(nE)$ |
| **for** each of n cells **do** | $O(n|\text{Min\_covers}|4^{E/4}2^{E}E^{E+2}(h!)^{h+1}) \approx O(nh^{4h^2})$ |
|    Generate_cell_layouts(n, HV); | |
| Find_optimal_cell_array; | $O(nE^2)$ |
| **end**; {HRM-TrailTrace} | |

**Fig. 6.12.** Worst-case time complexities of components of the *HRM-TrailTrace* algorithm.

Since the height $h$ of a functional cell is bounded by four, the time complexity of the algorithm, when restricted to the domain of practical circuits, becomes independent of $h$, and therefore of $E$. Hence, the algorithm's time complexity $O(nh^{4h^2})$ reduces to $O(n)$. Consequently, *optimal* layouts of large cell arrays consisting of an arbitrary number of functional cells can be generated in *linear* time.

## 6.3 EXPERIMENTAL RESULTS

We have implemented *HRM-TrailTrace* in a 9000-line Pascal program. Now we present the results of running this program on several commercial circuits, and various circuits found in the layout literature.

### 6.3.1 Commercial Circuits

We have generated layouts in the style of Fig. 6.1 for some commercial CMOS circuits found in IC manufacturers' data books [Si86]. The results are summarized in Fig. 6.13. This list includes a representative sample of the circuits for which logic circuits are provided and that do not contain transmission gates. The transistor circuits used by *HRM-TrailTrace* are approximated by mapping logic gates into standard transistor realizations. The actual layouts for these circuits are not available from the manufacturers; the main reason for generating their layouts is to demonstrate that the execution time of *HRM-TrailTrace* is reasonable for medium to large-sized commercial circuits. Note that the circuits of Fig. 6.1 have up to 44 functional cells and up to 190 transistors. The execution time of *HRM-TrailTrace* depends largely on the particular cells that constitute an array. The same dependency is observed in *HR-TrailTrace*. As shown in Chapter 5, *HR-TrailTrace* generates layouts for 98% of all practical-sized cells in less than 20 minutes on a MicroVAX II, but a few require more time. Typically, *HRM-TrailTrace* processes these cells in roughly the same amount of time as *HR-TrailTrace*. Ten of the twelve circuits took 15 minutes or less, with the other two requiring a few hours. This range of times on a small computer is well within the scope of feasible computation effort specified by our hypothesis in Chapter 1. As noted already, relatively inexpensive workstations are presently available that are roughly two orders of magnitude faster than the MicroVAX II; on such computers, even the worst-case run times of this algorithm should be completely satisfactory for practical applications.

| IC part number | Function | Number of functional cells | Number of transistors | Execution time |
|---|---|---|---|---|
| n/a | Gated SR latch | 2 | 16 | 8 sec. |
| n/a | Full adder | 2 | 26 | 1 min. |
| n/a | 2-bit comparator | 5 | 40 | 30 sec. |
| 7476 | JK flip-flop | 4 | 40 | 1 min. |
| 74183 | Carry-save full adder | 5 | 42 | 2 min. |
| 7482 | 2-bit full adder | 7 | 66 | 10 min. |
| 74182 | Carry look-ahead generator | 7 | 88 | 78 min. |
| 74148 | 8-to-3 priority encoder | 22 | 114 | 150 min. |
| 74248 | BCD-to-seven segment decoder | 15 | 138 | 11 min. |
| 74181 | 4-bit ALU | 44 | 142 | 15 min. |
| 74280 | 9-bit parity generator | 17 | 144 | 2 min. |
| 74259 | 8-bit addressable latch | 21 | 190 | 2 min. |

**Fig. 6.13.** Execution times of *HRM-TrailTrace* on some commercial circuits.

## 6.3.2  Published Layouts

We have collected several layout examples published in the literature on one-dimensional cell arrays. From these, we have selected arrays of dual series-parallel functional cells that do not contain transmission gates or dynamic logic. Most have a layout style similar to ours, which facilitates the comparison. The cell-array area is computed by counting the number of polysilicon columns and diffusion gaps to determine cell width, and the number of metal rows to determine height. This allows us to compare quite directly the size of the various layouts independent of the specific IC technology used. The results of the comparison are given in Fig. 6.14.

| Reference | Figure | Function | Published layout size | | HRM-TT layout size | | % area impro-vement | Run time (s) |
|---|---|---|---|---|---|---|---|---|
| | | | Height in rows $(U,C,L)$ | Width in columns | Height in rows $(U,C,L)$ | Width in columns | | |
| [BA88] | 7-7 | XOR | (3,1,8) | 6 | (5,1,2) | 5 | 80 | 2 |
| [BA88] | 8-5 | Full adder | (6,1,6) | 18 | (7,1,3) | 14 | 52 | 18 |
| [KW85] | 13 | MSFF | (6,4,3) | 16 | (6,1,3) | 15 | 39 | 22 |
| [Hi85] | 3 | 2-bit adder | (3,7,3) | 38 | (8,1,3) | 31 | 33 | 44 |
| [Ma85] | 14 | Random | (5,1,3) | 16 | (4,1,3) | 14 | 29 | 12 |
| [Uv81] | 2c | XOR | (4,1,2) | 6 | (5,1,2) | 5 | 5 | 2 |

**Fig. 6.14.** Comparison of *HRM-TrailTrace's* layouts to published cell-array layouts.

The area differences between the published layouts and our optimal layouts can be quite large. For instance, the layout of the 2-bit full adder generated by the Sc2 program [Hi85] is 33% larger than that of *HRM-TrailTrace*. Another example is the master-slave D flip-flop from TOPOLOGIZER [KW85], shown in Fig. 2.7, whose layout is 39% larger than our layout in Fig. 6.15. Other differences are even greater, and highlight the importance of evaluating heuristics by using optimal layouts as standards of comparison. Without the performance guarantees that approximation algorithms offer, unbounded heuristics can produce results arbitrarily far from the optimum [HS78]. Even comparing heuristics to manual layouts can give false confidence that near-optimality is achieved; the flip-flop layout of TOPOLOGIZER is only 3% larger than the manual design to which it is compared in [KW85], yet it is 39% larger than our optimal layout. These results underline the value of conducting a broad search over a slightly constrained space, compared to searching heuristically over a less constrained space. Often very general layout tools cannot evaluate many different transistor placements because the associated routing problem is too complicated. We can examine a larger number of placements with *HRM-TrailTrace* because we can evaluate the associated simple routing problem quickly.

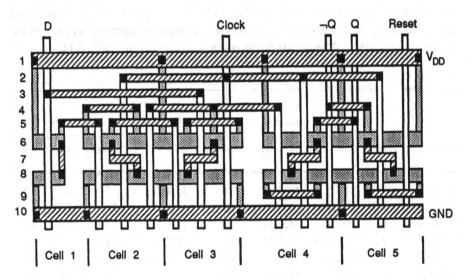

**Fig. 6.15.** A cell-array layout by *HRM-TrailTrace* for the master-slave D flip-flop of [KW85].

# CHAPTER VII
# CONCLUSIONS

In this chapter, we discuss the major contributions of our research, and its practical applications and extensions.

## 7.1  CONTRIBUTIONS

The main contributions of this work are threefold. First, we have developed a complete theory for exact layout minimization of one-dimensional functional cells and arrays of such cells. Second, based on this theory, we have developed exact feasible algorithmic solutions to each of the eight fundamental area-minimization problems defined for functional cell layout in Chapter 2, of which previously four had been unsolved and two only partially solved. Third, using our algorithms, we have demonstrated that exact layout minimization is not only feasible, but offers significant area reduction over present heuristic and limited-scope algorithmic methods.

**Minimization Theory.**  We have developed a method that exactly characterizes the eight layout minimization problems of Fig. 2.4 and forms the basis for practical algorithms to solve these problems.  A primary component is the theory of trail covering for dual series-parallel and planar graphs, which is based upon the concepts of cover types and cover concatenation operations corresponding to the graph composition process.  We have shown how complete sets of trail covers lead to exact layout width and height minimization. We have also characterized the composition of nonseries-parallel graphs; this characterization allows the theory of series-parallel circuits to be extended to nonseries-parallel planar circuits for the first time.  Another important component of our theory is an efficient method for reordering series-parallel graphs; this method enables our algorithms, unlike all previous layout methods,  to cover optimally all practical circuits.

153

**Exact Minimization Algorithms.** Our *TrailTrace* and *P-TrailTrace* algorithms solve the W problem, the exact width-minimization problem without circuit reordering, for dual series-parallel and planar circuits, respectively. *R-TrailTrace* solves the WR problem, the exact width-minimization problem with reordering; this is the original problem proposed by Uehara and vanCleemput [Uv81]. *HR-TrailTrace* solves the WH and WHR problems, the single-cell, exact width- and height-minimization problems, without and with reordering, respectively. We have also shown how *TrailTrace* and *R-TrailTrace* can solve the WM and WRM problems, exact width-minimization for multiple cells, without and with reordering, respectively. *HRM-TrailTrace* solves the WHM and WHRM problems, the exact width- and height-minimization problems for cell arrays. To solve the above height minimization problems, we defined complete layout models for single- and multiple-cell layouts in which minimum height, in addition to minimum width, can be measured and achieved in practice.

We have shown the computational feasibility of these algorithms by programming them and applying them to all practical-sized, single-cell planar circuits, and many multiple-cell circuits. In this way, we have demonstrated that exact layout area minimization of such circuits is indeed computationally feasible, according to the hypothesis we proposed in Chapter 1.

**Significant Area Reduction.** We have also demonstrated that exact minimization offers significant area reduction over heuristic and limited-scope algorithmic methods, thus validating the area claims of our hypothesis. We conducted a comprehensive experimental study in which we applied programs implementing our algorithms to all practical-sized CMOS single-cell circuits. We demonstrated that exact reordering can reduce layout width of a single cell by over 20%, and that exact height minimization with reordering can reduce a layout's area by more than 80%. By classifying these circuits according to their essential properties, this study also showed that the limited-scope algorithms of Nair and Madsen can only claim optimality on 56% of these circuits, and those of Lefebvre and Chan only on 18% of them [NB85, Ma89, LC89]. By generating all optimal layouts of these circuits, we showed that the heuristic method of Uehara and vanCleemput misses the optimal solution for 51% of the practical circuits. We have also shown that our algorithms offer significant area savings by applying them to virtually all single- and multiple-cell circuits published in one-dimensional functional cell layout generation papers and comparing our layouts to those of other methods. Although the single-cell examples show considerable area reductions, the savings are particularly dramatic on the multiple-cell examples, with improvements as high as 80%.

## 7.2   PRACTICAL APPLICATIONS AND EXTENSIONS

The most direct application of our algorithms is to incorporate them into CAD tools for generating layouts in the design of complex ICs. Once a large logic function is partitioned into cells, these algorithms can be used to generate very area-efficient layouts for the cells that are not speed-critical since, in their present form, they minimize area but do not directly address circuit performance. As our experimental studies have shown, our approach offer significant area savings over present heuristic methods. These savings can provide companies with a competitive advantage over those that use less efficient manual design methods, or methods that do not achieve minimum area. Such savings in area can either increase the amount of circuitry on a chip, or reduce chip size to lower IC manufacturing cost. In addition, we believe it is straightforward to extend our methods to generate high performance layouts by either limiting circuit reordering to fast circuits only, or by choosing the fastest circuit of minimum size. Likewise, our approach is amenable to circuits with transistors of various sizes, to multiple metal routing layers, and to the use of multiple contacts [Ma91].

Our algorithms can also be used to evaluate heuristic layout generation methods that use the functional cell style. In industry, heuristic layout methods are sometimes used, either alone or as parts of larger CAD systems. It would be very useful to generate the optimal area layouts using our methods and compare them to those of alternative heuristics. In this way, one could determine not only which heuristic is best, but also how far from optimal its layouts are.

In conclusion, we have demonstrated, based on a rigorous layout minimization theory, that exact width and height minimization of practical-sized CMOS circuits is feasible and offers significant area advantages over existing methods.

# BIBLIOGRAPHY

[AT90]   AT&T Corp., *0.9μm CMOS Cell Library: Standard Cells and Functional Blocks*, 1990.

[BA88]   D. G. Baltus and J. Allen, "SOLO: a generator of efficient layouts from optimized MOS circuit schematics," *Proc. 25th Design Automation Conf.*, pp. 445-452, 1988.

[Ba89]   R. Bar-Yehuda et al., "Depth-first-search and dynamic programming algorithms for efficient CMOS cell generation," *IEEE Trans. Computer-Aided Design*, vol. CAD-8, pp. 737-743, July 1989.

[Bo79]   B. Bollobas, *Graph Theory: An Introductory Course*, New York, Springer-Verlag, 1979.

[BS90]   M. L. Brady and M. Sarrafzadeh, "Stretching a knock-knee layout for multilayer wiring," *IEEE Trans. Computer*, vol. C-39, pp.148-151, Jan. 1990.

[BP87]   W. Bridgewater and R. Pokala, "Cell synthesis for ASICs," *Proc. Custom Integrated Circuits Conf.*, pp. 371-374, 1987.

[Br77]   R. A. Brualdi, *Introductory Combinatorics*, New York, North-Holland, 1977.

[Ca90]   B. S. Carlson et al., "Dual independent cell generation," *Proc. IEEE Int. Symp. Circuits and Syst.*, pp. 1636-1639, 1990.

[Ch90]   S. Chakravarty et al., "On optimizing nMOS and dynamic CMOS functional cells," *Proc. IEEE Int. Symp. Circuits and Syst.*, pp. 1701-1704, 1990.

[Ch91]    S. Chakravarty et al., "Minimum area layout of series-parallel transistor networks is NP-hard," *IEEE Trans. Computer-Aided Design*, vol. CAD-10, pp. 943-949, July 1991.

[CC89]    C. C. Chen and S. Chow, "The layout synthesizer: an automatic netlist-to-layout system," *Proc. 26th Design Automation Conf.*, pp. 232-238, 1989.

[CH88]    C. Y. R. Chen and C. Y. Hou, "A new layout optimization methodology for CMOS complex gates," *Proc. Int. Conf. on Computer-Aided Design*, pp. 368-371, Nov.1988.

[CK89]    H. Y. Chen and S. M. Kang, "Performance driven cell generator for dynamic CMOS circuits," *Proc. IEEE Int. Symp. Circuits and Syst.*, pp. 1883-1886, 1989.

[De87]    E. Detjens et al., "Technology mapping in MIS," *Proc. Int. Conf. on Computer-Aided Design*, pp. 116-119, Nov. 1987.

[Do89]    A. Domic, et al., "CLEO: a CMOS layout generator," *Proc. Int. Conf. on Computer-Aided Design*, pp. 340-343, Nov.1989.

[Du65]    R. J. Duffin, "Topology of series-parallel networks, *Journal of Mathematical Analysis and Application*, vol. 10, pp. 303-318, 1965.

[Er85]    C.K. Erdelyi, "Random logic design utilizing single-ended cascode voltage switch circuits in nMOS," *IEEE Journal of Solid-State Circuits*, vol. SC-20, pp. 591-594, April, 1985.

[GJ79]    M. R. Garey and D. S. Johnson, *Computers and Intractability: A Guide to the Theory of NP-Completeness*, New York, W. H. Freeman, 1979.

[Ga82]    D. D. Gajski, "The structure of a silicon compiler," *Proc. Int. Conf. Computer Design*, pp. 272-276, 1982.

[Ga85]    D. D. Gajski, "Silicon compilation," *VLSI Systems Design*, pp.48-64, Nov. 1985.

[GD85]   L. A. Glasser and D. W. Dobberpuhl, *The Design and Analysis of VLSI Circuits*, Reading, MA, Addison-Wesley, 1985.

[Gr76]   D. L. Greer, "An associative logic matrix," *IEEE Journal of Solid-State Circuits*, vol. SC-11, pp. 679-691, Oct. 1976.

[Ha69]   F. Harary, *Graph Theory*, Reading, MA, Addison-Wesley, 1969.

[Ha65]   M. A. Harrison, *Introduction to Switching and Automata Theory*, New York, McGraw-Hill, 1965.

[He87]   H. Heeb, "A rule-based system for polycell generation," *Fast-Prototyping of VLSI*, (G. Saucier, E. Read and J. Trilhe, eds.) Amsterdam, Elsevier North-Holland, pp.109-116, 1987.

[HF87]   H. Heeb and W. Fichtner, "GRAPES: A module generator based on graph planarity," *Proc. Int. Conf. on Computer-Aided Design*, pp. 428-431, Nov. 1987.

[Hi85]   D. Hill, "Sc2: A hybrid automatic layout system," *Proc. Int. Conf. on Computer-Aided Design*, pp. 172-174, Nov. 1985.

[Hi89]   D. Hill et al., *Algorithms and Techniques for VLSI Layout Synthesis*, Boston, Kluwer Academic Publishers, 1989.

[HS78]   E. Horowitz and S. Sahni, *Fundamentals of Computer Algorithms*, Potomac, MD, Computer Science Press, 1978.

[HH90]   Y. Hsieh, C. Hwang, Y. Lin and Y. Hsu, "LIB: a cell layout generator," *Proc. 27th Design Automation Conf.*, pp. 474-479, 1990.

[HS88]   Y. M. Huang and M. Sarrafzadeh, "Parallel algorithms for minimum dual-cover with application to CMOS layout," *Proc. Int. Conf. on Parallel Processing*, pp. 26-33, Aug. 1988.

[Hu85]   A. Hui et al., "A 4.1K gates double metal HCMOS sea of gates," *Proc. Custom Integrated Circuits Conf.*, pp. 15-17, 1985.

[HH89]    C. Hwang, Y. Hsieh, Y. Lin, and Y. Hsu, "An optimal transistor-chaining algorithm for CMOS cell layout," *Proc. Int. Conf. on Computer-Aided Design*, pp. 344-347, Nov. 1989.

[IM71]    T. Ibaraki and S. Muroga, "Synthesis of networks with a minimum number of negative gates," *IEEE Trans. Computer*, vol. C-20, pp.49-58, Jan. 1971.

[Ju90]    K. Just et al., "Palace: a layout generator for SCVS logic blocks," *Proc. 27th Design Automation Conf.*, pp. 468-473, 1990.

[Ki84]    J. H. Kim et al., "Exploiting domain knowledge in IC cell layout," *IEEE Design & Test*, pp. 52-64, Aug. 1984.

[Ko88]    P. Kollaritsch et al., "CLAY: a malleable-cell multi-cell transistor matrix approach for CMOS layout synthesis," *Proc. Int. Conf. on Computer-Aided Design*, pp. 142-145, 1988.

[KW84]    P. W. Kollaritsch and N. H. E. Weste, "A rule-based symbolic layout expert," *VLSI Design*, pp. 62-66, Aug. 1984.

[KW85]    P. W. Kollaritsch and N. H. E. Weste, "TOPOLOGIZER: An expert system translator of transistor connectivity to symbolic cell layout," *IEEE Journal of Solid-State Circuits*, vol. SC-20, pp. 799-804, June 1985.

[KK88]    Y. Kwon and C. Kyung, "A fast heuristic for optimal CMOS functional cell layout generation," *Proc. IEEE Int. Symp. Circuits and Syst.*, pp.2423-2426, 1988.

[LC89]    M. Lefebvre and C. Chan, "Optimal ordering of gate signals in CMOS complex gates," *IEEE Custom Integrated Circuits Conf.*, pp. 17.5.1-17.5.4, 1989.

[LC90]    M. Lefebvre, C. Chan and G. Martin, "Transistor placement and interconnect algorithms for leaf cell synthesis," *IEEE European Design Automation Conf.*, pp. 119-123, 1990.

[Le90]    T. Lengauer, *Combinatorial Algorithms for Integrated Circuit Layout*, Chichester, Wiley, 1990.

[LM88]   T. Lengauer and R. Müller, "Linear algorithms for optimizing the layout of dynamic CMOS cells," *IEEE Trans. Circuits and Syst.*, vol. CAS-35, pp. 279-285, March 1988.

[LD89]   I. Lin, D. Du and S. Yen, "Gate matrix layout synthesis with two-dimensional folding," *Proc. 26th Design Automation Conf.*, pp. 37-42, 1989.

[LG88]   Y. Lin and D. Gajski, "LES: A layout expert system," *IEEE Trans. Computer-Aided Design*, vol. CAD-7, pp. 868-876, Aug 1988.

[LN90]   P. Lin and K. Nakajima "A linear time algorithm for optimal CMOS functional cell layouts," *Proc. Int. Conf. Computer Design*, pp. 449-453, 1990.

[LT79]   R. J. Lipton and R. E. Tarjan, "A separator theorem for planar graphs," *SIAM Journal of Applied Mathematics*, pp. 177-189, April 1979.

[LL80]   A. D. Lopez and H. S. Law, "A dense gate matrix layout method for MOS VLSI," *IEEE Trans. Electron Devices*, vol. ED-27, pp. 1671-1675 Aug. 1980.

[LG84]   C. Lursinsap and D. Gajski, "Cell compilation with constraints," *Proc. 21st Design Automation Conf.*, pp. 103-108, 1984.

[LG85]   C. Lursinsap and D. Gajski, "Methods of cell compilation with constraints," *Proc. 22nd Design Automation Conf.*, pp. 303-307, 1985.

[Ma89]   J. Madsen, "A new approach to optimal cell synthesis," *Proc. Int. Conf. Computer-Aided Design*, pp. 336-339, 1989.

[MD88]   F. Mailhot and G. DeMicheli, "Automatic layout and optimization of static CMOS cells," *Proc. Int. Conf. on Computer Design*, pp. 180-185, Oct. 1988.

[Ma85]   T. Mano et al., "OCCAM to CMOS," in *Computer Hardware Description Languages and their Applications*, (C.J. Koomen and T. Moto-oka, eds.) Amsterdam, Elsevier North-Holland, pp. 381-390, 1985.

[Mi86]    H. Miyashita et al., "An automatic cell pattern generation system for CMOS transistor-pair array LSI," *Integration*, vol. 4, pp. 115-133, 1986.

[Ma91]    R. L. Maziasz, "Exact layout area minimization of static CMOS cells," Ph.D. Dissertation, Computer Science and Engineering Program, University of Michigan, 1991.

[MH87]    R. L. Maziasz and J. P. Hayes, "Layout optimization of CMOS functional cells," *Proc. 24th Design Automation Conf.*, pp. 544-551, 1987.

[MH90]    R. L. Maziasz and J. P. Hayes, "Layout optimization of static CMOS functional cells," *IEEE Trans. Computer-Aided Design*, vol. CAD-9, pp. 708-719, July 1990.

[MH91]    R. L. Maziasz and J. P. Hayes, "Exact width and height minimization of CMOS cells," *Proc. 28th Design Automation Conf.*, pp. 487-493, 1991.

[MO85]    C. T. McMullen and R. H. J. M. Otten, "Layout compilation of linear transistor arrays, *Proc. IEEE Int. Symp. Circuits and Syst.*, pp. 5-7, 19895

[MO88]    C. T. McMullen and R. H. J. M. Otten, "Minimum length linear arrays in MOS," *Proc. IEEE Int. Symp. Circuits and Syst.*, pp.1783-1786, 1988.

[MC80]    C. Mead and L. Conway, *Introduction to VLSI Systems*, Addison-Wesley, Reading, MA, 1980.

[NB85]    R. Nair, A. Bruss and J. Reif, "Linear time algorithms for optimal CMOS layout," *VLSI: Algorithms and Architectures*, (P. Bertolazzi and F. Luccio, eds.) Amsterdam, Elsevier North-Holland, pp. 327-338, 1985.

[NJ85]    T. Ng and S. L. Johnsson, "Generation of layouts from MOS circuit schematics: a graph theoretic approach," *Proc. 22nd Design Automation Conf.*, pp. 39-45, 1985.

[NH85]    J. T. Nogatch and T. Hedges, "Automated design of CMOS leaf cells,"*VLSI Systems Design*, pp. 66-78, Nov 1985.

[OL89]    C. Ong, J. Li and C. Lo, "GENAC: an automatic cell synthesis tool," *Proc. 26th Design Automation Conf.*, pp. 239-244, 1989.

[Pi83]    C. Piguet, "Design methodology for full custom CMOS microcomputers," *Integration*, pp. 335-350, Dec 1983.

[Po89]    C. J. Poirier, "Excellerator: custom CMOS leaf cell layout generator," *IEEE Trans. Computer-Aided Design*, vol. CAD-8, pp. 744-755, July 1989.

[Ra80]    F. R. Ramsay, "Automation of design for uncommitted logic arrays," *Proc. 17th Design Automation Conf.*, pp. 100-107, 1980.

[RN77]    E. Reingold, J. Nievergelt and N. Deo, *Combinatorial Algorithms: Theory and Practice*, Englewood Cliffs, NJ, Prentice Hall, 1977.

[RH85]    C. Rowen and J. L. Hennessy, "SWAMI: A flexible logic implementation system," *Proc. 22nd Design Automation Conf.*, pp. 169-175, 1985.

[Ru87]    S. M. Rubin, *Computer Aids for VLSI Design*, Reading, MA, Addison-Wesley, 1987.

[Sa89]    G. Saucier et al., "A channelless layout for multilevel synthesis with compiled cells," *Proc. Int. Conf. on Computer Design*, pp. 35-38, Oct. 1989.

[ST85]    G. Saucier and G Thuau, "Systematic and optimized layout of MOS cells," *Proc. 22nd Design Automation Conf.*, pp. 53-61, 1985.

[SM83]    K. H. Schmidt and K. D. Mueller-Glasser, "NMOS dense gate matrix VLSI layout," *IEEE Journal of Solid-State Circuits*, vol. SC-18, pp. 157-159, April 1983.

[SK88]    T. Shiple, P. Kollaritsch and J. Allen, "Area metrics for transistor placement," *Int. Conf. on Computer Design*, pp. 428-433, Oct. 1988.

[Sh88]    Y. Shiraishi et al., "A high packing density module generator for CMOS logic cells," *Proc. 25th Design Automation Conf.*, pp. 439-444, 1988.

[Si86]    Signetics Corp., *High-speed CMOS Data Manual*, 1986.

[Sm83]    K. F. Smith, "Design of regular arrays using CMOS in PPL," *Proc. Int. Conf. Computer Design* , pp. 158-161, 1983.

[SN88]    A. Stauffer and R. Nair, "Optimal CMOS cell transistor placement: a relaxation approach," *Proc. Int. Conf. Computer-Aided Design*, pp. 364-367, Nov.1988.

[Su89]    P. K. Sun, "CETUS: a versatile custom cell synthesizer," *Proc. Int. Conf. on Computer-Aided Design*, pp. 348-351, 1989.

[Ta82]    T. Takamizawa et al., "Linear-time computability of combinatorial problems on series-parallel graphs," *Journal of the ACM*, vol. 29, pp. 623-641, July 1982.

[TI86]    Texas Instruments, *2-μm CMOS Standard Cell Data Book*, 1986.

[TS88]    G. Thuau and G. Saucier, "Optimized layout of MOS cells," *IEEE Trans. Computer*, vol. C-37, pp.79-87, Jan. 1988.

[Uv78]    T. Uehara and W. M. vanCleemput, "Optimal layout of CMOS functional arrays," *Proc. 16th Design Automation Conf.*, pp. 287-289, 1978.

[Uv81]    T. Uehara and W. M. vanCleemput, "Optimal layout of CMOS functional arrays," *IEEE Trans. Computer*, vol. C-30, pp. 305-312, May 1981.

[Ul84]    J. D. Ullman, *Computational Aspects of VLSI*, Rockville, MD, Computer Science Press, 1984.

[We67]    A. Weinberger, "Large scale integration of MOS complex logic," *IEEE Journal of Solid-State Circuits*, vol. SC-2, pp. 182-190, Dec. 1967.

[WE85]    N. H. E. Weste and K. Eshraghian, *Principles of CMOS VLSI Design*, Reading, MA, Addison-Wesley, 1985.

[Wi87]    S. Wimer et al., "Optimal chaining of CMOS transistors in a functional cell," *IEEE Trans. on Computer-Aided Design*, vol. CAD-6, pp. 795-801, Sept. 1987.

[Wl82]    O. Wing, "Automated gate matrix layout," *Proc. IEEE Int. Symp. Circuits and Syst.*, pp. 681-685, 1982.

[Wi83]    O. Wing, "Interval-graph-based circuit layout," *Proc. Int. Conf. Computer-Aided Design* , pp. 84-85, 1983.

[Wi85]    O. Wing, "Gate matrix layout," *Proc. IEEE Int. Symp. Circuits and Syst.*, pp. 199-202, 1985.

[Wo83]    W. Wolf et al., "Dumbo, a schematic-to-layout compiler," *Proc. Third Caltech Conf. on VLSI*, pp. 379-393, 1983.

[YH85]    E. J. Yoffa and P. S. Hague, "ACORN: A local customization approach to DCVS physical design," *Proc. 22nd Design Automation Conf.*, pp. 32-38, 1985.

[YK85]    M. Yu and W. J. Kubitz, "Linear-time heuristics for cell layout synthesis under constraints," *Proc. Int. Conf. Computer-Aided Design* , pp. 64-66, 1985.

# INDEX